Fundamentals of Human Biology and Health

Third Edition

By Heather Murdock
San Francisco State University

cognella™
San Diego, CA

Bassim Hamadeh, CEO and Publisher
Christopher Foster, General Vice President
Michael Simpson, Vice President of Acquisitions
Jessica Knott, Managing Editor
Kevin Fahey, Cognella Marketing Manager
Jess Busch, Senior Graphic Designer
Stephanie Sandler, Licensing Associate
Sean Adams, Editorial Assistant

First published in the United States of America in 2013 by Cognella, Inc.

Trademark Notice: Product or corporate names may be trademarks or registered trademarks, and are used only for identification and explanation without intent to infringe.

Printed in the United States of America

ISBN: 978-1-62131-263-5

www.cognella.com 800.200.3908

This is dedicated to my daughters, Ella and Gillie. I love their curiosity, fascination, and sense of humor when it comes to the human body!

Acknowledgments

Many thanks to my wonderful family and friends for their encouragement and support during this project, especially Gillian, Ella and Paul. I'm also very appreciative of my SFSU students, past and present, for all of their excellent questions and comments during class. I have incorporated many of these into this text. It's much more fun to teach when students are engaged and interested in what they are learning. I have always been very impressed with the students at San Francisco State University in this regard. I have a handful of students who really helped me by providing their typed class notes so that I would know what I actually ended up writing on the board. Contributions from Joe Bernisky, Diana Rosas, Cheyenne Snavely, and Kristina Stuckenbrock are all very much appreciated. Thank you to Dr. Alan Salamy and Allen Hatchett for reading this manual and providing feedback and suggestions. Thank you also to everyone that helped put this reader together from University Readers, Inc., especially my excellent copy-editor Sharon Hermann. And most of all, thank you to my mom and dad for always enthusiastically encouraging me throughout my education, I'm happy to pass along what I've learned to others.

Contents

Preface

This textbook is based on my lecture notes from past semesters of teaching human biology lectures and general biology laboratories at San Francisco State University. I have taken the information from a variety of sources listed in the bibliography including books, magazine and newspaper articles, health journals, and internet sources. This is not a complete picture of the human body and how it works. It is more a summary of the basic forms and functions of different systems in our incredible bodies, along with some topics that I think students might find interesting and relevant. I have written this book because I think there needs to be a more affordable option than the $150 dollar price tag usually associated with human biology textbooks. We are all dealing with budget issues these days, especially in relation to education. I realize many students reading this are not biology majors and may never have to study biology again, but I hope I can relay the basics so that you will have a solid understanding of how human cells, tissues, organs, and organ systems operate, as well as knowledge that will allow you to make good choices about what you do with your own bodies. I will be focusing on basic health issues, including nutrition, exercise, and disease prevention, as well as reinforcing why we need to help one another and the planet.

Because this is really just skimming the surface of every topic, I encourage you to read other books to supplement your knowledge, especially for the subjects that you find interesting or that you don't fully understand just from my simplified coverage. It's also great to seek other sources for more pictures and diagrams because visuals are very helpful when learning a new subject. (You can pretty much Google anything these days!) This textbook is a work in progress, and I will be adding to it for future semesters. The study of biology is constantly evolving and I learn new information every time I teach a class. I find that my students have a lot to teach me as well, so let's learn together and hopefully have some fun in the process.

Heather

Chapter 1

Introduction to Biology and the Diversity of Life

Biology is the study of living organisms and their characteristics. I think everyone should have a basic understanding of this science. Biology relates to everything about us: who we are, where we came from, what we are composed of, how we function, how organisms are interrelated, how the choices we make affect our own bodies, as well as those around us and our planet. I hope some of the topics covered in this book will inspire you to lead healthy lifestyles and to respect other organisms and ecosystems. Learning about basic biology may also inspire you to delve into one of the numerous biological fields. There are new discoveries all of the time and it's a very exciting subject to study.

Biology has led to the prevention and treatment of so many diseases, with lifesaving advances such as vaccines, antibiotics, chemotherapy, dialysis, anaesthesia, and surgical procedures. We also understand how organisms have evolved over time and can compare DNA and morphological features of different organisms and fossils to see how species are related. We can even map the entire human genetic code now, an international scientific project that took 13 years to complete! There are so many interesting careers associated with biological studies, such as zoologist, doctor, dentist, teacher, pharmacist, physical therapist, ophthalmologist, game warden, landscape architect, veterinarian, geneticist, paramedic, conservation biologist, environmental consultant, pathologist, dietitian, nurse, lab technician, dermatologist, medical illustrator, oceanographer … the list goes on and on.

If this hasn't grabbed your interest, then maybe some of the amazing statistics we'll be covering about the human body will fascinate you. We are so efficient at balancing our temperature, water, and electrolyte levels; using nutrients; making energy; eliminating waste; fighting disease; sorting and storing data in our brain; reproducing; and responding to our environment inside and out! We will be discussing the following topics throughout the book, but here are a few fun facts to throw out at parties to impress your friends.

There are 2 million species classified on our planet today but it has been extrapolated that there have been anywhere from 5 to 100 million living and fossil species from when life began (3.8 billion years ago!).

Modern humans have been around for 195,000 years.

Human, chimpanzee, and bonobo DNA sequences are about 98% the same!

Humans have 46 chromosomes, which consist of approximately 25,000 genes coding for all of the traits that make us who we are.

An adult human is made up of 80–120 trillion cells, 25% of our which are red blood cells.

There are over 200 different types of cells, all made from stem cells in our bone marrow. Most are microscopic (you could line up 2,000 blood cells across your thumb) but we have larger cells too, like the neurons in nerves that stretch over 3 feet long!

Speaking of nerves, if you lined up all of the peripheral nerves in your body (the nerves in areas other than the brain and spinal cord) they would be 93,000 miles long!

We can send 300 messages per second with our nerves and the impulses can travel 395 feet per second (over 150 miles per hour). We generate enough electricity with the billion neurons in our brain to power a light bulb (or turn on a flashlight)!

We have 10 million chemoreceptors in our nose to help us recognize 10,000 odors. Odor is the strongest trigger for memories.

We have 2.5 million sweat glands, on average; sweat is one of the main ways we lose water from our bodies, but it's also very important to help us cool down when we get too hot.

We have trillions of skin cells covering our bodies (3–10 trillion depending on the size of the body) and we shed about a million skin cells a day! This makes up most of the dust around your house! (Yikes)

The average adult head has 120,000 hairs; however, averages tend to vary by pigmentation: 140,000 for blondes and more like 90,000 for redheads. We lose about 100 hairs per day.

Half of the human body tissue is composed of muscle. (We have three types—skeletal, smooth, and cardiac muscles.) There are 640 skeletal muscles, the largest of which is the gluteus maximus (which helps form the buttocks and gives it a rounded shape). All of the skeletal muscles working together in an adult body could lift 11 tons, which is equivalent to 4 SUVs!

The number of bones in your body depends on how old you are. As babies, we start out with 300 bones. They are soft bones at this stage, made of cartilage and easily pliable, which helps us to get out of the birth canal. They have all hardened completely into bone by the time we are 14 months old. As we develop, certain bones fuse together, and by the time we are 20 years old, most people have 206 bones. (However, some people have an extra rib and some have an extra bone in the arch of the foot, so the total can be 207 or 208 depending on your genetics.) Our body is composed of 18% bone.

Our heart is a truly amazing organ made up of cardiac muscle. It can pump 5.6 liters of blood continuously all day (the equivalent of cycling 2,000 gallons a day). Each blood cell travels a complete circuit around the pulmonary and systemic circulatory systems within 60 seconds. The heart averages 100,000 beats per day. That can mean 2.5 billion beats in a 70-year life. If you added up all of our blood vessels (arteries, arterioles, capillaries, venuoles, and veins) they would reach about 64,000 miles. (That is over twice the circumference of the Earth!)

Our kidneys cycle 45–50 gallons of fluid every day by filtering our blood 30 times a day through the one million nephrons we have in each kidney. This process is incredibly important to help us remove our wastes (such as urea, uric acid, and creatinine), as well as excess water and salts, and in the process, reabsorb everything that we need to keep in our blood (like nutrients, electrolytes, and water).

The adult stomach can expand to hold 2 liters of food. Everything we eat is broken down into smaller particles throughout the digestive tract in an incredibly efficient manner. Most of what we need to keep is reabsorbed into our blood from the small intestines, which are almost 20 feet long. The small intestine has tiny projections called microvilli to increase the surface area of the tubules into 2,800 square feet (the size of a tennis court). Are you impressed yet?

I am constantly amazed and fascinated by the human body. We often take our bodies for granted when everything is working perfectly. On the other hand, we can generally detect the slightest imbalance since it often causes many affects on how we function and feel. I hope reading this book will help to spark an appreciation for all of the trillions of parts we are composed of. They are working together in such an intricate manner to make us the incredible machines that we are.

This book focuses on humans, but we will discuss other types of organisms along the way since we are all interrelated. Living organisms are incredible on so many levels. What does it mean when we say "living organisms"? Seems pretty basic, but you need to know the definition of life in order to classify whether something is living or not. For example, viruses, although they can spread and cause great harm, are not actually classified as "living organisms." They do not possess all of the characteristics that the scientific community has come up with to define life.

Main Traits of Life:

1. Living things are made up of **cells**. Organisms can be uni- (one-celled) or multicellular (many-celled).
2. Living things **respond to stimuli** in their environment, through chemical receptors or some sort of nervous system—from the very basic to our incredibly complex nervous system.
3. Living things increase in size and/or number or cells (they **grow and develop**).

4. Living things **reproduce** new organisms of their own species and pass on hereditary material (in the form of DNA). Reproduction can be asexual (cloning themselves) or sexual (reproducing with a partner, which increases genetic variability). Reproduction is necessary for the survival of the species.

5. Living things use an **energy** source (from food) to fuel functions. Energy use is also known as **metabolism**, where chemical reactions in the body break down nutrients and build compounds needed for life.

6. Living things maintain an internal environment that is favorable to cell function. The physical and chemical environment inside the body must be kept within certain limits that can support life. For example, we need to keep the pH of our blood around 7.4; if blood is too acidic or too basic, our cells can't function—death can result from these conditions. "Staying the same" is also called "**homeostasis**."

7. Living things may **adapt to environmental changes** resulting in an increased ability to reproduce. Darwin's Natural Selection theory states that the organisms that are best adapted to their environment will pass on the most genes. Species will change over time in the process we call Evolution. Organisms have been evolving on Earth for the past 3.8 billion years. Before we can understand ourselves, we need to take a look at what has gone on before humans even inhabited the planet.

Life on Earth

Earth formed 4.6 billion years ago (bya)

Life began roughly 3.8 bya (primitive anaerobic bacteria cells evolved, possibly from protobionts (organic molecules in a membrane-like structure).

Primordial atmospheric gases formed these organic molecules.

We are still unclear of the process.

Photosynthesis by early bacteria 3 bya produced <u>oxygen</u>,

an important gas for other species to evolve.

Eukaryotic cells (more complex than prokaryotic cells) evolved 2 bya.

Complex **multicellular life** evolved 1 bya (colonies of cells, like early **protista**).

Fungi are hard to pinpoint in the evolutionary record, but some studies believe early species appeared around 800 million years ago.

Simple **Animals**, like sponges, evolved 665 million years ago from a type of protozoa called choanoflagellate.

Complex animals are in the fossil record around 550 mya (such as echinoderms, early chordates, arthropods, and mollusks, all of which were still in the sea; there were no terrestrial organisms yet).

Fish evolved from early chordates close to 500 mya.

Fungi started to colonize land around this same time.

Plants evolved from the protist algae Charophyta around 475 mya.

Insects and **gymnosperm plants** (plants with "naked" seeds) came along 400 mya.

Dinosaurs and other **reptiles** evolved 300 mya, after the **amphibians** (360 mya).

Mammals evolved 200 mya (small insectivorous egg-laying mammals came first and the placental and marsupial mammals evolved 70 million years later).

Birds evolved from dinosaurs 150 mya.

Angiosperms (plants with enclosed seeds and flowers)came 130 mya.

Non-avian dinosaurs became extinct 65 mya;

mammals radiated everywhere and evolved even more;

the first primates showed up in the fossil records around this time.

25 mya ancestors of apes and humans diverged from ancestors of old-world monkeys.

Distant bipedal ancestors of man occurred 5 mya.

2.5 mya the genus homo is found in the fossil evidence.

200,000 ya humans started looking like they look today.

25,000 ya, extinction of Neanderthals.

Population isolation, natural selection, and sexual selection caused different genetic traits in various populations, forming what we now call racial differences (skin color, hair color, hair texture, eye shape, and body stature).

(Dates prior to 1 billion years ago are speculative.)

Diversity of Life

There are so many different kinds of species on our planet. We've classified only about two million to date, but it's been extrapolated that there have been anywhere from five to 100 million living and fossil species on Earth! It's important to understand some basics about all of the major groups of organisms since we are all interconnected.

Scientists now group organisms into **three Domains**. Before 1990 they were usually grouped into **five Kingdoms**; since I'm old school, I still like the kingdom method, but I'll give you both here:

The three domains: Archaea, Bacteria, and Eukarya.

The five kingdoms: Monera, Protista, Plantae, Fungi, and Animalia.

Here are some of the defining characteristics of the domains and kingdoms.

<u>Three Domains:</u>

1. **Archaea**—unicellular prokaryotic organisms. (Prokaryotes don't have membrane-bound organelles, they just have free DNA, cytoplasm, and microtubules making flagella or cilia on the outside for movement.) Archaea usually live in extreme conditions without oxygen, in places like swamps, volcano vents, landfills, salt water, and acidic water. They are VERY similar to the bacteria domain in form, but their gene sequences are different enough to classify them in a separate domain. (Although, when classifying organisms based on the kingdom method, I would still put the Archaea and Bacteria together.)

2. **Bacteria**—also unicellular and prokaryotic and are found EVERYWHERE! They are the types found in food, our bodies, soil, on plants, etc., etc. They are mostly **decomposers** and get food from outside sources—breaking down plant and animal matter and absorbing the nutrients into their cells—but some are **producers** and are able to make their own food through photosynthesis (using sunlight) or chemosynthesis (using chemicals). (This is also referred to as "autotrophic" = "self-feed.") Cyanobacteria (blue-green algae) are incredibly important since they started producing their own food 3 bya and in the process made oxygen, which led the way for most other species to evolve. Bacteria are the first kingdom in the old classification system.

3. **Eukarya**—the third and last domain is extremely diverse, and holds the four other kingdoms—protista, plants, animals, and fungi—all of which have eukaryotic cells. Eukaryotic cells differ from prokaryotic cells in their complexity and the fact that they have discrete membrane-bound organelles performing different functions within the cells. We will cover the parts of the cell in Chapter 3, but some organelles that you have probably heard of are the mitochondria (to make energy), endoplasmic reticulum (to help make proteins), Golgi bodies (shipping and processing), and the nuclei (location of DNA synthesis and transcription). Some eukarya are autotrophic and can make their own food, such as the plants and various protista species. Many eukarya are called "heterotrophic" (other-feed) since they cannot make their own food and must get it from another source, either as a decomposer or a **consumer**.

Now let's break the classification down into the **Five Kingdoms**:

1. The most primitive kingdom is called **Monera**, which are all prokaryotic cells (see the Archaea and Bacteria description above). Over 10,000 species are known in this Kingdom (but there could be millions that have not been discovered yet). For all of the rough estimates of species known per kingdom, just realize that new species are being discovered all the time (about 15,000 per year) and that every source has different estimates, so it's hard to pinpoint the number of species for any of the kingdoms!

The **eukaryotic kingdoms** are

2. **Protista****: Many different kinds of protista exist (roughly 200,000 known species); some are producers, consumers, or decomposers; some are multi-, some unicellular; some don't have cell walls, and some do (cellulose or chitin). Some can move via flagella, cilia, or pseudopodia. They are sometimes classified according to whether they are animal-like (move and ingest other organisms), plant-like (can photosynthesize), or fungus-like (produce spores). Examples include paramecia, algae, slime molds, and parasites like plasmodium (causes malaria). An example of a protist that is very important for all life on Earth are the diatoms (a type of alga) that live in the sea—they make roughly one-third of the oxygen we all breathe! Plants evolved from a species of protista in the alga category and animals evolved from a protist that was similar to a sponge-like animal.

3. **Plants**: There are over 300,000 known species and all are producers, which means they make their own food through photosynthesis (even carnivorous plants can photosynthesize as well). They make oxygen during photosynthesis and use carbon dioxide, are multi-cellular, and have cell walls made of cellulose. Plants, along with algae, provide most of the food and oxygen for everyone else on the planet, directly or indirectly. In fact, for humans, 90% of our world's food comes from only 20 plant species! Of the 27 tons of oxygen necessary for all life on Earth, 80% is produced by algae and 20% is produced by plants. Plants are also critical for medicine (80% comes from plants), building material, paper, clothes, clean air and water, climate, erosion and sediment control, habitat for other organisms, recreation, and aesthetics.

4. **Fungi**: Most of the 100,000 known species are multi-cellular, although one of our favorite fungi is unicellular (yeast: important for breads, beer, wine, etc.). Fungi have cell walls of chitin and they are usually decomposers, and some are parasites. They cannot move on their own. They are very important because they help recycle nutrients back into the soil. This kingdom boasts the largest species in the world with the "humongous fungus" in Oregon that is 3.5 miles wide and covers 2,000+ square miles!

5. **Animals**: The largest kingdom, with over 1.3 million known species (over 1 million of which are insects versus only about 5,500 mammals). Animals are all multi-cellular organisms, have no cell wall, and are consumers because they can't make their own food and must get it from an outside source. (Herbivores eat only from the plant, fungi, and/or protist groups; carnivores eat only animals; and omnivores will eat certain organisms from all four eukaryotic categories.) Animals require oxygen for cellular processes and produce carbon dioxide in the process. They can move by using cilia, flagella, or muscle systems. There are two major groups of animals: <u>invertebrates</u> (animals without backbones make up 98% of the animal kingdom) such as annelids (worms, leeches, etc.), arthropods (insects, spiders, crustaceans, etc.), echinoderms (starfish, sea urchins, etc.), and mollusks (snails, clams, octopi, etc.); and the <u>vertebrates</u> (animals with a backbone are the remaining 2%, roughly 43,000 species) such as fish, amphibians, reptiles, birds, and mammals.

** There are really more than five kingdoms since the bacteria are divided into two kingdoms and the protista have been broken into several kingdoms, but for our purposes we will just refer to five kingdoms to simplify things.

The Inclusive Hierarchical System of Classification

The study of biology is also about ordering the natural world from macro to micro or vice versa, in order to understand connections between organisms.

Taxonomy assigns organisms a name, versus **phylogeny**, which goes beyond naming to understand the relationships between organisms.

Carl Linnaeus developed an amazing scientific classification system in 1758 that is still used today (with slight modifications).

This system of classification is hierarchical in that the taxonomic categories form groups within groups, assigned by phylogenetic relationships. Higher categories contain greater numbers of species and have broader definitions:

The major taxonomic categories used in biology are

Kingdom;
    Phylum or Division (for plants);
        Class;
            Order;
                Family;
                    Genus;
                        Species.

When I was in school I was taught a phrase to remember the order:
Kings **P**lay **C**hess **O**n **F**ine **G**lass **S**tools

Now that Domains have been added, I have heard students say,
 "Dumb kids play catch on freeway go splat!"
You can make up your own way of remembering the hierarchical order to classify organisms.

Some basic definitions of some of these groups that we talk about regularly:

Taxon (**taxa**): Refers to a taxonomic group at any level.
Genus: A group of species related by common descent and the species within a genus share certain derived features.
Species: A group of organisms that can interbreed and produce fertile offspring.
Binomial name: Name for a species that consists of a genus name and a species epithet.

Genus and species name are always italicized or underlined, and always Latinized (ex: *Homo sapiens* = "wise man"). Every species of organism has one and only one scientific name, governed by the **International Codes of Nomenclature**. Common names are useful but sometimes confusing, especially when different languages have different common names.

Humans are Eukarya, Animalia, Chordata (subphyla vertebrata), Mammalia, Primates, Hominidae, *Homo sapiens*.

Looking at some of the features that put us in these taxa:
 Eukarya because we have complex membrane-bound cells; **Animals** since we are all multicellular, consumers (heterotrophic by ingestions), we have no cell walls, are motile (can move), and our embryos pass through a blastula stage. **Chordata** means we share features with all the chordates, like notochord, dorsal hollow nerve tube, post-anal tail (in utero), and pharynx with gill slits (in utero), and also have vertebra and teeth. We are **mammals** with hair, mammary glands, and three middle ear bones (and a placenta since we're placental mammals); we are **primates** with forward-facing eyes, color vision, opposable thumbs, and many facial expressions (as opposed to some animals who have only one). **Hominids** have even more complex social organization and brain development, longer parental care, larger body and brain size, sexual dimorphism, and 32 teeth. And finally, as ***Homo sapiens*** (humans), we have great manual dexterity with extremely developed nervous and muscular systems, erect posture, a highly complex brain with sophisticated language skills, self-awareness, ability to plan for the future, etc. (Just to clarify the connection between humans, apes, and monkeys, we are not descended directly from apes or monkeys; rather, we have common ancestors.) The human–ape line diverged from the old-world monkey line 25 million years ago, and then the humans and apes diverged again 5 to 8 million years ago. Human, chimpanzee, and bonobo DNA are all around 98% the same due to this relationship.

Okay, now that we have a basic understanding of the species on Earth and where we fit in, let's talk about how we are related in terms of energy and nutrients.

Energy Flow—everything is related, and energy and nutrients are recycled in the process:

The ultimate source of energy is the SUN, which provides the energy for plants and other producers to make sugars (glucose) via **photosynthesis**, one of the most important biological reactions on Earth. Photosynthesis is the process by which producers make their own food by converting carbon dioxide and water into glucose with the help of the Sun's energy. Another very important product of the reaction is oxygen, needed for all life on Earth.

$$\textbf{Photosynthesis: } CO_2 + H_2O \rightarrow \text{Sun's energy} \rightarrow \text{Glucose } (C_6H_{12}O_6) + O_2$$

In order for photosynthesis to occur in plants, they must take in enough water and CO_2. Water is taken in through the roots and transported through xylem to the leaves, where photosynthesis occurs. The CO_2 enters the leaves through pores called **stomata**; the pores are opened and closed by cells on each side called **guard cells**. When they open the CO_2 can come in—usually during the day when there is enough light for photosynthesis, and then they close at night. This helps to keep water in, since it evaporates out when the stomata are open. Sunlight is needed to provide the energy necessary to power the reaction and is captured by pigments in the chloroplasts of the leaves. The primary pigments are chlorophyll a and b, which absorb mostly red and blue wavelengths of light and reflect green (this is why most plants are green!). The sugar made by photosynthesis in the leaves is then transported to other parts of the plant for cellular respiration and storage by another type of vascular tissue called phloem. The oxygen made by photosynthesis exits out of the stomata. This is a very simplified explanation of photosynthesis.

Meanwhile, all organisms (including plants) use the glucose made during photosynthesis to synthesize their own energy in the biological process called **cellular respiration**. (Since producers are the only organisms that can make glucose by photosynthesis, all other organisms must get the glucose from producers, one way or another.) Cellular respiration converts glucose and oxygen into carbon dioxide, water, and **adenosine triphosphate (ATP)**. ATP is an energy-storage molecule that releases energy when it is broken down into adenosine diphosphate. Every living organism performs cellular respiration in the mitochondria of their cells to create energy. Even producers need to make their own energy despite being able to use energy from the Sun to synthesize their food. Energy is necessary for every living organism to run cellular processes.

$$\textbf{Cellular respiration: Glucose } (C_6H_{12}O_6) + O_2 \rightarrow CO_2 + H_2O + ATP$$

Notice that the photosynthesis and cellular respiration equations are opposite of one another. The amount of oxygen that the producers make (roughly 27 tons per year) is almost exactly equal to the amount of oxygen that all of the organisms on Earth take in for cellular respiration. This means the amount of oxygen in the atmosphere stays constant. (Carbon dioxide levels, however, have not stayed constant due to increased production caused by humans. The final chapter will address this issue when we cover pollution and climate change.)

CO_2 is toxic and we can't handle TOO much water in our system either, so we need to get rid of the byproducts of cellular respiration. They both diffuse out of our cells into our circulatory system; we release CO_2 when we breathe out and release H_2O in our urine, with our breath, and through our skin. (Water makes up at least 60% of our tissues, however, so we do need to keep a lot of water in our systems for many of our chemical reactions and to move substances around in our body.)

CO_2 is toxic because it combines with H_2O to form carbonic acid (H_2CO_3), which can then break apart to form hydrogen ions (H^+) and carbonate (HCO_3). The buildup of acids in our cells can dissolve cell membranes and the cellular contents spill into the body. We have a complex system to make sure that we don't build up too much acid in our cells. (The fact that you can't hold your breath for more than a few minutes is your body's way of making sure you get rid of your CO_2!)

Our bodies don't make or destroy chemicals, but rather reorganize to build different compounds or expel them. As I already pointed out, these simplified equations of **cellular respiration** and **photosynthesis** are opposite of one another—the plants need CO_2 to make their sugar, oxygen is a byproduct that we need to make our energy, and our byproduct is CO_2, which is used by plants. (They are also using CO_2 in our atmosphere created as a byproduct of the combustion of fossil fuels or the burning of vegetable matter, among other chemical processes. Carbon dioxide is also emitted from volcanoes, hot springs, and geysers, and also comes from breaking down carbonate rocks.) Plants go through cellular respiration as well as photosynthesis (they are both happening simultaneously during the day). So they are making oxygen as well as carbon dioxide at the same time.

Animals can't make their own sugar, so they rely on plants as the glucose source for cellular respiration. Herbivores and omnivores eat the plants to get the energy from the starch that the plants are making via photosynthesis; other animals (carnivores and omnivores) eat other animals who have already consumed plants and are now storing the sugar as glycogen. Meanwhile, the decomposers like fungi and bacteria decompose all of the dead plant, animal, etc., matter and use the sugar for cellular respiration to make energy, and in the process they recycle the nutrients like nitrogen and carbon to the soil to be used again by plants. Everything is interconnected. We'll look at how affecting one part of the chain of life affects others when we get to the last portion of the book and talk about ecosystems and conservation. But for now you should know the basic equations of photosynthesis and cellular respiration.

In order to classify life on Earth and understand how different processes work, scientists have universal guidelines that they follow when researching a topic. It has taken thousands of years of curious people to have the knowledge we now have about our planet and its inhabitants, and scientists keep discovering new things every day. Here are the fundamental steps used when performing scientific research.

The Steps of the Scientific Method:

1. **Observation**. Pay attention to what's going on around you. It can be on a macro- or microscopic level. Ask questions about your observations.
2. **Hypothesis**. Form a tentative explanation to your question(s) called a hypothesis (an educated guess), based on your observations. You can make predictions based on this hypothesis to be <u>tested</u>. (If your hypothesis cannot be tested, it is not valid for a scientific investigation since you cannot prove or disprove your hypothesis.)
3. **Experiment**. Design an experiment to test your hypothesis; make sure to have enough <u>replication</u> and a <u>control</u> (something that differs from your experimental treatment by one variable).
4. **Analyze results**. See if there is statistically significant data to either accept or reject your hypothesis.
5. **Conclusion**. Provide a detailed explanation of what your data mean in terms of your hypothesis. If your hypothesis is proven to be incorrect, you can start all over again with a new hypothesis. Every experiment is helpful!
6. **Share** with the rest of your class, the scientific community, or the public at large so that others can build on the new knowledge and review your work. (Otherwise science wouldn't move forward as fast as it does, since many scientists would be researching the same topics over and over, and without peer review, the experiment might not be considered credible.)

Chapter 2

Basic Chemistry and Organic Molecules

Okay, now that we've covered some basics in terms of science in general, we're ready to dive into a little chemistry, necessary for understanding the processes going on in the human body all of the time. We'll be looking at how small particles, like ions, affect our cells, organs, and organ systems. As we near the end of our course, we'll move up to how humans as a whole affect ecosystems. Here's some vocabulary to order and define matter. We will start from the smallest particle of matter and move up from there.

Hierarchical Order of Matter and Life

Quark—scientists have discovered particles even smaller than subatomic particles!

Subatomic Particles—protons (+), neutrons (neutral), and electrons (-) are within an atom.

Atom—a basic unit of matter composed of protons and neutrons inside a nucleus with electrons circling around the nucleus. The number of protons in the atom determines the <u>atomic number</u>. The atomic number classifies the atoms into chemical elements. (So basically different types of atoms are called elements.)

Element—consists of only one type of atom and is categorized by its atomic number. Each atom has a different number of protons. For example, Hydrogen is an atom with one proton and therefore its atomic number is 1. Oxygen is an atom with 8 protons and is number 8 on the periodic table. The periodic table tells us how many protons are in each element (**atomic number**) and the weight of the element (**atomic mass**: the weight of the protons and neutrons combined). The mass of the electrons in an atom is negligible and therefore is not included in the atomic mass.

Ion—a charged atom or molecule with an unequal number of protons and electrons. If there are more protons in the ion the charge is positive, if there are more electrons it is a negatively charged ion.

Electrolyte—ions that dissolve in water and make a substance electrically conductive. Examples are sodium (Na^+), calcium (Ca^+), chloride (Cl^-), and potassium (K^+). These are very important for biological processes like muscle contractions and nerve impulses.

Molecule—a unit made up of two or more atoms of the same or different elements (when different elements combine they are called **chemical compounds**); ex: H_2O is two hydrogen and one oxygen.

Organelle—"little organ" = intracellular compartment surrounded by a protective membrane (ex: mitochondria, nucleus, etc.); biological molecules form organelles.

Cell—smallest unit with the ability to live and reproduce.

Tissue—made up of cells and substances that are combined to perform a specific function.

Organ—composed of tissue combined to perform a specialized function.

Organ System—two or more organs interacting to perform particular functions.

Multicellular Organism—an individual organism made up of cells, organized as tissues, organs, and organ systems.

Species—a group of reproducing organisms that can produce fertile offspring of the same kind (ex: ligers are not a true species since they cannot make a new liger by reproducing together—the only way to make a liger is to mate a lion and a tiger).

Population—group of organisms of the same species living in the same area.

Community—populations of different species living in the same area.

Ecosystem—the community and its environment (organisms living together plus the abiotic factors such as water, air, and soil).

Biome—large-scale communities classified by main vegetation, climate (amount of precipitation and temperature), and combination of organisms. Examples include desert, savanna, taiga, tundra, grassland, and tropical or temperate forests.

Biosphere—regions of Earth's crust, water, and atmosphere inhabited by living things (made up of different biomes).

Let's look at some of our chemistry terms in more detail. Hopefully this will help you to understand how molecules come together and break apart when we're discussing different cellular processes, as well as why ions and pH affect our systems as much as they do. To understand how the human body works, you need some very basic chemistry. I am definitely not a chemistry buff so we will really only cover the very minimum that you need for future biology topics in this text.

Matter is anything that takes up space. Matter can be in the form of a solid, liquid, or gas. The units of matter are **atoms**.

Elements consists of one type of atom. Elements cannot be broken down into another type of substance. The periodic table has 118 elements, although most of these elements are rare. There are only 11 main elements in living organisms: O, H, N, C, Ca, Na, K, Cl, Mg, P, and S. (There are trace amounts of other elements but they are very minuscule.) There are only 10 main elements in the Earth's crust, ocean, and atmosphere: O, Si, Al, Fe, Ca, Na, K, Mg, H, and Ti. After you take into consideration the overlapping main elements from living organisms and the environment, there are really only 15 common elements out of the 118 discovered to date. Now let's look at some properties of elements. Looking at the chart you will see Helium's <u>atomic number</u> is 2, which means that helium has two protons, and its <u>atomic mass</u> is 4, which is the sum of the masses of **protons** and **neutrons** (each proton or neutron has one Dalton; He has two of each, for a mass of 4). The mass of **electrons** is so small it's not considered in the atomic mass.

All atoms have the same basic structure: the atom has a nucleus with protons and neutrons in the center and electrons orbit the nucleus. The protons have a positive charge and if it has neutrons (which are neutral) they are in a 1:1 ratio. The only element without any neutrons is Hydrogen (the atomic mass of H is only 1). If the neutrons are not in the same proportion as the proton (can be more or less), it is called an **isotope**. Ex: Regular carbon has six protons and six neutrons, but the isotope carbon-14 has six protons and eight neutrons. Many elements have isotopes, with the same number of protons but a different number of neutrons. All isotopes of a particular element are chemically identical but have different masses.

Tiny electrons (negative charge) orbit the nucleus at high speeds in **valence shells**. Atoms have an equal number of protons and electrons and therefore are neutral. If an atom gains or loses an electron, it becomes a charged atom called an **ion**, which has a charge corresponding to the number of electrons lost or gained. Electrons are lost or gained from the atom's outer valence shell. Atoms with many electrons have multiple valence shells and each shell can hold a limited number of electrons in the outer shell. The 1st shell can hold two electrons, maximum; all of the following shells (2nd shell, 3rd shell, 4th shell, etc.) can hold up to eight electrons in the outer shell. (However, starting with the 3rd valence level, shells inside the outer shell can start building up more than 8 electrons, but the outside shell will never have more than 8 electrons.)

Chemical properties of an atom depend on the number of electrons in the outermost shell and an atom is stable if the outer shell is complete. Atoms with a filled outer shell are **inert** or unreactive elements and are called **rare** or **noble gases** (they are on the right edge of the periodic table). These atoms with no "vacancies" usually don't take part in chemical reactions. Other elements with their outer shells missing electrons can share, gain, or lose electrons to achieve the stability of a filled outer valence. Elements that have either one or seven electrons in the outer valence shell are particularly reactive (which we will see when we discuss Sodium Chloride).

There are different ways elements react and bond with one another. These interactions result in atoms staying together to form **molecules** through these chemical bonds. Molecules can be the same or different atoms joined together. When elements combine to form substances consisting of two or more different elements, they are called **chemical compounds** such as water (H_2O). Compounds consist of two or more elements in proportions that never vary; for example, every H_2O molecule compound has one oxygen atom bonded with two H atoms.

There are four elements that make up 96% of the chemical composition of living matter: **C**arbon, **O**xygen, **H**ydrogen, and **N**itrogen. There are small quantities of other elements in humans as well, such as of calcium, phosphorus, sulfur, sodium, chlorine, magnesium, and trace elements such as iron, iodine, and selenium, but we are going to concentrate on the four main elements since they make up the bulk of our bodies. We are composed of about 60% to 75% H_2O, and the other "dry weight" is mostly organic building blocks such as proteins, carbohydrates, lipids, and nucleic acids. We will discuss these more in depth at the end of this chapter but for now, let's look at some of the ways the elements are bonded together to form these molecules in our bodies.

Sharing at least one pair of electrons is known as **covalent bonding**: ex: H-H (H_2) uses a single covalent bond (sharing one pair of electrons). Two oxygen atoms (O_2) are bonded together with a double covalent bond (O=O), which means that they share two pairs of electrons. This is an even stronger bond than a single covalent bond. (The more pairs of electrons shared, the more stable the bond becomes.) Covalent compounds are usually liquids or gases at room temperature, like H, N, O, and all of the halogens (the row to the left of the noble gases—F, Cl, Br, I, At, Uus). Covalent bonds are found in CO_2, O_2, H_2O, and glucose, all critical for the human body.

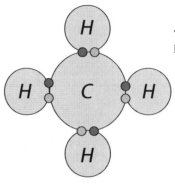

The elements in this molecule of CH_4 are held together by covalent bonds.

○ Hydrogen electrons
● Carbon electrons

Atoms with unfilled orbitals in their outermost shell tend to interact with other unstable atoms; as I said before, there are different ways they can interact. We just covered sharing of electrons; another way is by transferring electrons from one atom to another. **Ionic bonds** form when electrons are transferred from one element to another and the elements become ions (charged atoms). Opposite ions attract to form ionic bonds. Sodium chloride is an example of a chemical compound held together by an ionic bond. Sodium (Na) has 11 protons and 11 electrons (the outer valence shell has one electron). When Na associates with Chlorine (Cl), which has 17 protons and 17 electrons (with 7 electrons in the outer valence shell), the Cl, which is more electronegative, strips one electron away from the sodium turning it into a positive ion (cation) (Na^+). The extra electron on the Cl makes it into a negative ion (anion) (Cl-), and the opposites attract and make NaCl with an ionic bond—a very weak bond that can be dissolved in water. (NaCl, sodium chloride, is also known as table salt, which is made up of thousands of sodium and chloride elements ionically bonded together to form salt crystals.) Molecules formed by ionic bonds are generally all solids at room temperature, as opposed to covalent bonds, which can be gases, liquids, or solids.

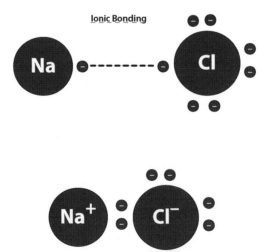

Ionic Bonding

Other weak bonds are **hydrogen bonds**, also vital to biology. There is a slight positive charge on hydrogen atoms that are covalently bonded to a more electronegative element, and this slight charge can form weak attractive bonds with adjacent slightly negative atoms. The hydrogen bond is much weaker than covalent or ionic bonds, but it does cause attractions between nearby molecules. Hydrogen bonds exist in all life forms since they are the bonds that join two strands of DNA together in the double helix configuration. The bonds are easily broken during DNA replication and then are formed again once the process is completed. The bonds are also present in joining water molecules together as well as forming the shapes of protein molecules.

Bonds do more than just hold elements/molecules together; some store energy when bonds are formed (like in the case of making sugars), and these are called **endothermic reactions**. Other bonds release energy (**exothermic reaction**) when molecules are broken apart. In an exothermic reaction, energy is released when the bond is broken.

Chemical reactions involve making or breaking chemical bonds and thus changing the composition of matter. Enzymes (which are proteins) help to speed up the rate of reactions.

$$\text{Ex: } H_2O_2 \rightarrow H_2O + O \text{ using peroxidase (enzyme)}$$

In this example, the <u>substrate</u> hydrogen peroxide (H_2O_2), which is toxic, is converted to the non-toxic <u>products</u> water and oxygen with the help of the <u>enzyme</u> peroxidase. Chemical reactions like this are occurring in our bodies all the time. Usually an endothermic reaction is coupled with an exothermic reaction so that energy (in the form of ATP) is released from the exothermic reaction to run the endothermic reaction. Reactions can occur without enzymes as well, but they would take a lot longer. We will go over enzymes in more detail when we talk about proteins.

Water is the most abundant and important molecule on Earth (the Earth is 73% water) as well as in living organisms. It makes up between 60 and 75 percent of total body weight in animals (more for certain animals like jellyfish!) and even more for plants—up to ~98% for some plants! Humans are composed of about 70% water. Our bodies need water to perform many basic functions like digestion, excretion, respiration, and circulation. Without enough water, the body becomes dehydrated, chemical reactions fail, and cells die. Too much water can cause our body to fail as well, and we will talk more about this when we discuss the urinary system, which helps control our water balance. Water helps to keep our body within a livable pH range as well. Our blood is mostly water and remains close to neutral with a pH of 7.4. If the blood becomes too acidic or basic we can die, so it's important for our body to constantly monitor and make adjustments when necessary to maintain homeostasis. An <u>acid</u> is a substance that produces hydrogen ions (H^+) in water and a <u>base</u> is a substance that produces hydroxide ions (OH^-) in water. Measuring the concentration of hydrogen ions in a solution gives you the amount of **potential Hydrogen**, also known as **pH**. The pH scale goes from 0 to 14 and anything under 7 is acidic, 7 is neutral, and above 7 is basic. Most of our biological fluids are between 6 and 8 on the pH scale with a few exceptions. (For example the gastric juice in the stomach is closer to 2, which is important for digestion and fighting pathogens that might enter with our food.)

Water is critical for chemical reactions as well. Water helps to build and break apart **organic molecules**, which make up the remaining 30% of the body weight in living organisms. <u>Organic</u> refers to molecules that contain carbon. Most organic molecules also contain at least one hydrogen atom as well. The **Four Main Organic Macromolecules** in all living things are Carbohydrates, Proteins, Lipids, and Nucleic Acids. Many of these organic molecules have the carbon and hydrogen linked in hydrocarbon chains, which are very stable structures. These chains can also have a functional group attached, which are particular atoms or clusters of atoms that are covalently bonded to carbon and can influence the chemical behavior of the macromolecule. One unit of macromolecule is a monomer; three or more monomers are a polymer. Water can be taken away or inserted into macromolecules in order to make or break molecules. These two very important opposite reactions are constantly occurring in our cells.

Dehydration Synthesis occurs when enzymes remove a hydroxyl (OH) from one molecule and a hydrogen atom (H) from another molecule, and the two molecules are then joined by a covalent bond (where the OH and H used to be) and the discarded OH and H form H_2O. (This is also called a condensation reaction.) This is a common building reaction to make molecules.

In order to break molecules apart, the opposite occurs: enzymes split molecules into two or more parts and then attach an OH group and an H atom from a molecule of H_2O to the exposed sites. Adding water helps break apart large polymers into smaller units, which can be used for building blocks (to make other macromolecules) or to make energy (ATP). This reaction of breaking molecules apart is called **Hydrolysis**.

Now let's discuss the **four main organic molecules** in more detail and go over their building blocks and functions.

1. **Carbohydrates** are the most abundant of the organic molecules. They always consist of Carbon, Hydrogen, and Oxygen atoms in a 1:2:1 ratio. Carbohydrates are sugars (saccharides) and are the BEST SOURCE OF ENERGY for humans.

Monosaccharides are **simple sugars**, which means they consist of only one monomer of saccharide. A critical monosaccharide is glucose, $C_6H_{12}O_6$. This sugar is made by producers during photosynthesis and is broken down (metabolized) by all organisms during cellular respiration to make energy. It is our main source of energy for all of our body cells. Another example of a simple sugar is fructose, found in many fruits, root vegetables, and honey. Galactose (in milk) and ribose and deoxyribose (in RNA and DNA) are also simple sugars.

Oligosaccharides are short chains of sugar units (two or more sugar monomers united by dehydration synthesis). Two simple sugars bonded together are called disaccharides. Examples of disaccharides include maltose = glucose + glucose, found in corn syrup; lactose = galactose + glucose, found in milk; and sucrose = glucose + fructose, found in most plants. Table sugar is sucrose crystallized from sugar cane or beets. Sucrose $(C_{12}H_{22}O_{11})$ was the main sweetener used in foods before high fructose corn syrup (HFCS) was introduced into many processed foods and sodas in the United States in the 1970s and 80s.

High Fructose Corn Syrup is not a naturally occurring carbohydrate (it was developed by scientists in the 1950s and 60s), but I would like to mention it here as well. It is basically a corn-based sweetener made up of glucose and fructose, and then modified by enzymes in order to convert the glucose into fructose to make it sweeter. HFCS is cheaper than sucrose and has become the number-one sweetener in the United States. Americans consume roughly 69 pounds of corn-based sweeteners per year (which is roughly 30 tsp per day, way over the recommended 6–9 tsp per day in a 2,000-calorie daily diet). We will discuss this further in the nutrition chapter but I just wanted to explain what it is here as well. HFCS may be one of the primary reasons obesity rates have risen so dramatically in the past 40 years.

Fructose is not metabolized the same way as glucose in our bodies. Instead of being used by cells to make energy, fructose is generally used by liver cells to make triglycerides (fat). We need a certain amount of fats as energy-storage reservoirs, insulation, and cushioning, but too much fat leads to obesity, type II diabetes, heart disease, and a whole variety of other health problems. Sweeteners in general are considered "empty calories" and should be limited since they do not add much nutritional value to our diet. The next type of carbohydrate, however, should make up the bulk of our diet since they provide us with the energy and building blocks we need to survive and they are also beneficial for the digestive process.

Polysaccharides are complex carbohydrates formed by chains of many sugar monomers joined by dehydration synthesis. The following main polysaccharides all consist of <u>glucose</u> monomers bonded together. Plants store glucose in the form of **STARCH** (used for energy), as well as **CELLULOSE** (in cell walls for structural support). Another polysaccharide used for structure is **CHITIN**, found in fungi and many animal exoskeletons and shells. This polysaccharide consists of cellulose and Nitrogen (for added strength). Animals store glucose for energy in **GLYCOGEN** (in muscles and liver cells).

These polysaccharides are all important to humans. We make our own polysaccharides in the form of glycogen in order to store energy. We cannot synthesize starch, cellulose, or chitin but we can use these polysaccharides for different purposes. We break down starch (from the plants or algae we consume) and glycogen (from animals we consume) into glucose monomers to be used in cellular respiration. Starch provides us with our primary source of energy. We can't digest cellulose (in the cell walls of plants and algae) or chitin (in fungal cell walls), but they are also important polysaccharides. Cellulose is especially important, and is the most prevalent organic molecule on Earth! Most plants are composed of at least 30% cellulose and some are much more (woods are around 40–50% and cotton contains 90% cellulose). Cellulose is in our building materials, clothes, paper products, and is a very important part of our diet. Even though we cannot metabolize cellulose, (since the bonds are much stronger than the starch and glycogen bonds), this polysaccharide, also known as "dietary fiber," helps with digestion. It softens our waste in the intestines and helps to move it out of the digestive tract. We will be talking about fiber again when we get to digestion and nutrition, but for now let's concentrate on our energy molecules.

In order to make energy through cellular respiration (occurring in the mitochondria of all our cells) we need to break down the carbohydrates to produce glucose monomers. The glucose then moves into the bloodstream, and the pancreas secretes the hormone **insulin** that helps our cells take up glucose faster. The brain uses roughly 65% of the glucose circulating in our blood at any given time. We transfer a lot of glucose to our muscle and liver cells as well—sometimes to make energy and sometimes for storage. Glucose can be converted to glycogen in the liver, which is one of our energy reserve molecules. If you need to make energy, and have run out of glucose in the blood, the pancreas secretes the hormone **glucagon** to convert the glycogen from the liver back to glucose, which can then be released into the blood to be used throughout the body. (Glycogen can also be made in smaller amounts in other organs and cells, for example in muscles, but when it is broken down it does not get released back to the bloodstream. Instead, it is used by the cells where it was stored.)

If we have plenty of glycogen, the liver can also convert excess sugars into fats (triglycerides), which are stored in adipose tissue in different locations around the body such as the buttocks, thighs, and abdomen. Our body can use these fats as an alternative to glucose to make energy when we run out of glucose and glycogen. In fact, fats are where we store most of our energy reserves (around 78%) versus the 1% in glycogen. Breaking down fatty-acid tails from triglycerides yields two times more energy than using glucose. We can also use proteins to make energy and this makes up the remaining 21% of our energy reserves. When the body uses proteins for energy, it breaks down the amino acid chains and converts them to fats, carbohydrates, or ATP, and the byproduct is ammonia. The ammonia is then converted to urea in the liver and is eventually excreted out with the urine. Here's a quick recap of what molecules we can use to make energy. We use glucose first when it is readily accessible in the bloodstream, (digested from the foods we consume). When we run out of glucose in the blood, we can metabolize the glycogen stored in the liver and muscle cells, and fats stored in adipose tissue or proteins located throughout the body.

2. **Proteins** are also incredibly important organic macromolecules. In fact, the combination of proteins that we have in our body really helps to determine who we are. Everyone's specific DNA sequence codes for which proteins are made in the cytoplasm of the cells. Proteins determine how our cells function, what we look like, how we behave, how our immune system responds to invaders, etc., etc. We all have a different DNA sequence (other than identical twins) so we all make different combinations of proteins. There are many proteins that everyone makes, although in slightly different amounts (like collagen, hemoglobin, actin, and myosin), and then there are others that help to create diversity within a species. For example, the different antibodies and antigens that we make create immunity diversity and give us different blood types. The different enzymes we make determine which eye or hair color we have. Most health disorders have to do with which proteins are made as well. For example, someone who is lactose intolerant is missing the protein that helps to break down the disaccharide lactose in dairy products. Proteins have many different functions, but first we will go over their composition.

Proteins account for more than 50% of the dry weight of cells. They contain all four of the main elements: H, C, O, and N, and are the most diverse of the macromolecules. The building blocks of proteins are a set of twenty **amino acids**. Each amino acid is composed of a carboxyl (COOH) + C + amino (NH_2) + H + R. (R signifies a side molecule that varies depending on which of the 20 amino acids you are talking about. The R molecule determines the chemical properties of how the amino acid behaves.) The twenty different amino acids are like letters in the alphabet. They can make millions of different proteins depending on how they are combined and arranged.

Amino acids are linked together by dehydration synthesis to form **peptide bonds**. Multiple amino acids linked together form a **polypeptide chain**. These chains coil and have a three-dimensional shape due to the hydrogen bonds' holding different parts of the chain together. The 3D structure of a protein determines its function. The sequence of the different amino acids is called the **primary structure** of the peptide or protein. There is also usually a **secondary**, **tertiary**, and **quaternary structure** as well. (The **quaternary structure** is the interaction between several chains of peptide bonds.) If temperature, pH, salt, or other environmental aspects change, the structure of the protein can be damaged. If the protein loses its 3D shape or the polypeptide chain is broken apart, it is "**denatured**," which means that it has lost its function. (Ex: high temperature will denature proteins because the heat will cause the hydrogen or peptide bonds to break.) Enzymes can renature if the suitable environment is restored in a test tube, but it may be more difficult in a living cell. Enzymes degrade over time naturally as well. Once they break apart, the amino acids are recycled within the cell to be used for more protein synthesis. If they are not needed for protein synthesis, they will be converted by the liver to ammonia and then urea, and subsequently excreted out in urine. Amino acids are not stored for later use like glucose.

Proteins have many important functions depending on their composition and shape. They can also be broken down for use as an energy source if there are no carbohydrates or lipids available. Here are some examples of different kinds of proteins according to function.

Structural proteins far outweigh the other four classes of proteins in terms of sheer mass within the body. Examples of structural proteins are collagen, the most common protein in the body, found mostly in different connective tissues; <u>myosin</u> and <u>actin</u>, the contraction proteins in muscle; and <u>keratin</u>, the building blocks of hair, skin, and nails.

Transport proteins can help move substrates in and out of cells (channel proteins and protein pumps) or carry substances with them in blood (examples: <u>hemoglobin</u> carries O_2 and <u>albumin</u> helps transport lipids), or muscle tissue (<u>myoglobin</u> transports O_2 to the muscle cells).

Regulatory proteins are generally very small hormones. (There are two classes of hormones; some are proteins and the others are lipids.) Hormones help to adjust cell activities. (Examples: <u>insulin</u> regulates glucose metabolism and <u>oxytocin</u> stimulates uterine contractions.)

Defense protein like <u>antibodies</u> can flag or kill invading pathogens. Other defense proteins are <u>antigens</u>, which help us to recognize self versus non-self.

Enzymes proteins probably provide the most important of the protein functions. Enzymes help run all of our chemical reactions (helping to make or break molecules). Examples: <u>salivary amylase</u> breaks down starches in our mouths during the digestive process. <u>DNA polymerase</u> helps to synthesize and repair DNA in all of our cells.

Enzymes are very important proteins to understand so I would like to explain how they work. Enzymes help to speed up reactions by bringing substrates together to form a product, or by helping to split molecules apart. Chemical reactions usually require energy, so as I've mentioned before, an exothermic reaction will often be coupled with an endothermic reaction so that there is ATP available as well. The enzymes are unchanged in the process (unless they are denatured by environmental conditions, or because they are worn down over time). Reactions are reversible depending on the concentration of the substrates and products. There are also "co-enzymes" that help enzymes with their job (usually vitamins or minerals) and there are "inhibitors" that impede enzyme functions, either by binding to the active site of an enzyme so that a substrate cannot bind (competitive inhibitor) or by binding to another site on the enzyme, which changes its shape so that it cannot accept the substrate any more (noncompetitive inhibitor). The shape of an enzyme determines which substrates it will accept and which products are made. Enzymes are extremely specific and efficient.

In an **Enzyme-Substrate Complex**, the substrate(s) attach in the active site of the enzyme (like a lock and key) and the product(s) leave after the reaction is complete. The enzyme-substrate complex only lasts a few milliseconds while the enzyme helps to convert the substrate(s) to the new product(s). The product(s) of the reaction then leave the active site to be used by the cell. The enzyme is ready to accept more substrate, and can typically act on more than 1,000 substrate molecules per second!

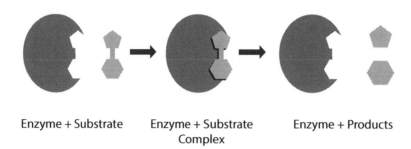

Enzyme + Substrate   Enzyme + Substrate   Enzyme + Products
Complex

Different environmental conditions affect enzymes' activity levels. Concentration of enzymes and substrates will affect reaction rate (adding more enzymes to a solution will speed up the reaction rate, since there are more sites for metabolic activity to occur). Each enzyme has an optimal temperature and pH condition. If the solution is too acidic or too basic, the reaction rate will decrease since the amount of salt and ions can denature enzymes. The optimal pH for many enzymes is around the pH of blood (~7.4); however, if enzymes are located in an acidic environment like the stomach, they tend to work best at a much lower pH.

Temperatures affect the speed of molecules and therefore the rate of collisions, as well as the binding qualities and shapes of molecules. Molecules move faster in warmer conditions, so usually as a solution is heated up, the molecules bump into one another more often and reactions run faster. Too much heat, however, breaks bonds and therefore can change an enzyme's shape. If this occurs, the enzyme is denatured and doesn't work at all, slowing the reaction rate as more and more enzymes denature. Boiling a substance will denature all of the enzymes and the reaction rate will cease altogether. Adding inhibitors to a solution would also decrease reaction rate, since these types of molecules inhibit enzymes from binding with their substrates.

3. Lipids are also important organic molecules with many different functions. There are four types of lipids: fats, phospholipids, steroids, and waxes. All lipids are non-polar hydrocarbons, insoluble in water and hydrophobic. They dissolve in similar non-polar solvents, however. Many contain one to three long chains of hydrocarbons called fatty acids, which are energy-storing molecules. The only lipids without any fatty-acid tails are the steroids.

Fats (such as triglycerides) account for 95% of the lipids in humans (The other 5% is a mixture of phospholipids and steroids with just a miniscule amount of wax.) Fats are composed of glycerol + one to three fatty-acid chains, linked by dehydration synthesis. Fat is good for storage of energy—it has two times the energy of polysaccharides (glycogen) when broken down. We store triglycerides as fat droplets in fat-storing tissue known as adipose cells. Fat is also important for cushioning and insulation. When we cover nutrition in chapter 11 we will also be talking about saturated versus unsaturated fats in our food. This refers to the molecular structure of the triglyceride fatty-acid chains. "Unsaturated fats," like oils, have at least one double covalent bond in a fatty-acid hydrocarbon chain. These are important fats that we need in our diet. If there are no double bonds in the hydrocarbon chains then every carbon is saturated with hydrogen and the triglyceride is said to be a saturated fat (like butter and other animal products). These are much harder for the body to break down, and may cause plaque buildup in blood vessels, leading to stroke and heart disease. Much of what we eat has a mixture of saturated and unsaturated fats. It's important to monitor how much fat we consume (especially the saturated variety). We will cover this in more depth when we get to nutrition and the heart.

Phospholipids are composed of two hydrophobic fatty-acid tails + glycerol + a hydrophilic head with a phosphate group. All of our cell membranes (enclosing our cells as well as our organelles) are made of phospholipid bilayers. The "bilayer" refers to the arrangement of the phospholipic molecules. The hydrophobic fatty-acid tails face one another toward the inside of the membrane, and the hydrophilic phosphate heads are on opposite ends of the bilayer, which means they would be on the inside and outside of the cell membrane. We will discuss this again in our next chapter when we go through the parts of the cell.

Steroids look different from fats and phospholipids because they do not have fatty-acid chains. Instead, steroids are four carbon rings fused together. An example of a very important steroid is cholesterol, which has several functions in our body. We need cholesterol to make bile in the liver (for lipid digestion) and some of our hormones are derived from cholesterol (testosterone, progesterone, estrogen, prostaglandins, and cortisol are all steroid hormones). Cholesterol gets a bad reputation, since everyone seems to be "watching their cholesterol levels," but it is an important compound in our body. In fact, it is needed by all of our cell membranes for flexibility and growth. We will talk more about "good" versus "bad" cholesterol in the nutrition chapter when we talk about cholesterol in foods. Another steroid that you have all heard about is the anabolic steroid that promotes growth in muscles. (We have natural steroids that do this, versus the synthetic ones that some athletes use.)

Waxes are composed of alcohols + a fatty-acid tail. We do not have much of this type of lipid in our body but our ears make wax for protection. How much wax you make is genetically determined. Plants make wax to cover their leaves (called a cuticle). This is both for protection from insects, bacteria, and fungi, as well as to keep moisture in and slow down transpiration rates (how much water evaporates out of the leaves). Wax production in plants also depends on where they live. Plants in very dry climates tend to make thicker cuticles.

4. Nucleic Acids are last in this list of organic molecules but certainly not the least. They are all critical organic molecules necessary for life. They are made up of <u>nucleotides</u>, which are formed from carbon, hydrogen, oxygen, nitrogen, and phosphorus. Nucleotides are made from a nitrogenous base (adenine, thymine, uracil, adenine, or guanine), a 5-carbon sugar (deoxyribose or ribose), and one to three phosphate groups. Which building blocks are used determines what kind of nucleic acid is made: DNA, RNA, or ATP.

DNA (Deoxyribonucleic Acid) programs all cell activity by encoding for proteins.
RNA (Ribonucleic Acid) has several forms and functions to aid with protein synthesis.
ATP (Adenosine Triphosphate) is the energy-storage molecule necessary for cellular processes.

Now let's cover the three types of nucleic acids individually and in more depth:

DNA (Deoxyribonucleic Acid) is made up of nucleotides, which consist of a 5-carbon sugar (deoxyribose) + a nitrogenous base (cytosine, thymine, adenine or guanine) + one phosphate molecule. The structure of DNA is described as a double helix because DNA is composed of two strands of polynucleotides in a spiral configuration. The two strands are held together by hydrogen bonds where the nitrogenous bases pair up. There are approximately 3.2 billion nucleotides in the DNA of every cell. The double helix nucleotides are arranged with proteins in units called chromosomes (when condensed) or chromatin (when not condensed), depending on whether the cell is undergoing cell division (chromosomes) or not dividing (chromatin). Humans have 46 chromosomes (23 pairs). The chromosomes range in length from 60 million nucleotides (the smallest chromosome is the Y chromosome, which determines "maleness") to 250 million nucleotides each on the first pair of chromosomes.

The sequence of nucleotides (and therefore nitrogenous bases) in our DNA determine our "genes." Genes code for different proteins that are made in the cytoplasm of our cells by the ribosomes. There are approximately 25,000 genes in humans (despite the fact that we have been able to map the human genome, we still don't know exactly how many genes we have). Pairs of chromosomes have different numbers of genes. For example, the first pair of homologous chromosomes has roughly 2,500 genes per chromosome versus pair #22, which contains only 500 genes each. Ninety-eight percent of DNA does not code for genes at all and is often referred to as "junk DNA." Some of this junk DNA, however, are important nucleotide sequences that regulate which proteins we make by turning gene expression on and off. We will discuss genes and chromosomes further when we get to cell division, reproduction, and genetics (chapters 16 and 17).

DNA is located in the nucleus of all eukaryotic cells and replicates itself during cell division. It cannot get out of the nucleus, but it can transfer the genetic information to another type of nucleic acid, RNA, which can exit the nucleus. The DNA nucleotide sequence (and thus its message) is transcribed to messenger RNA, which can then take the message to the cytoplasm to code for protein synthesis with the help of the ribosomes. In order to understand how DNA bonds in its double strand configuration, as well as how it replicates itself and transcribes the message to RNA, you need to know the specific nucleotide base pair rules:

There are only 4 types of nitrogenous base pairs in DNA or RNA molecules. They are not, however, the same 4 types; there is one substitution that makes DNA and RNA nucleotides different from one another. So there are 5 nitrogenous bases in all.

Nitrogenous Bases

Purines:	Pyrimidines:
Guanine (DNA & RNA)	Cytosine (DNA & RNA)
Adenine (DNA & RNA)	Thymine (DNA only)
	Uracil (RNA only)

In DNA, guanine always pairs with cytosine, and adenine always pairs with thymine, and vice versa: so if one strand was ATCGAATTGCG, the second strand that would line up and H-bond to the first would be TAGCTTAACGC. These base pair rules are slightly different when DNA is coding for RNA (DNA and RNA do not bond together like two strands of DNA do, but they still line up according to base pair rules when the DNA is coding to make an RNA strand. Remember that RNA has Uracil instead of Thymine, therefore a DNA adenine would code for RNA uracil, DNA Thymine codes for RNA adenine, and guanine and cytosine still code for each other.

DNA-DNA	DNA-RNA
A – T	A – U
C – G	C – G
G – C	G – C
T – A	T – A

RNA (Ribonucleic Acid) is a single strand of polynucleotides composed of a nitrogenous base (Cytosine, Uracil, Adenine, or Guanine) + the 5-carbon sugar ribose + a phosphate. RNA is transcribed (during a process called **transcription**) from the nucleotide language of DNA to the same nucleotide language of RNA (with the exception of uracil) and the base pair rules listed above. A strand of DNA with one chain sequence of ATCCG will code for an RNA strand of UAGGC.

DNA codes for three types of RNA.

Ribosomal RNA (**rRNA**) helps to build ribosomes, which are the organelles necessary for protein manufacturing.

Messenger RNA (**mRNA**) takes the DNA message out of the cells' nucleus into the cells' cytoplasm where the message is **translated** at the **ribosomes** to create **proteins** out of amino acid chains.

Transfer RNA (**tRNA**) also helps with translation by bringing amino acids to the messenger RNA in the protein-building process.

It is called translation because now the message is going from nucleotides to amino acids, which are different chemical "languages." Each sequence of **three nucleotides** codes for **one amino acid** (ex: UAC = Try, CAU = His, AAG = Lys). Amino acids linked together in a particular sequence and held together by peptide bonds form a **protein**. Each nucleotide sequence on a chromosome's DNA that codes for proteins is called a gene. (There are many sequences on each chromosome coding for many proteins.) The theory used to be that each gene coded for only one specific protein, but now we are learning that sometimes genes can code for more than one protein. With our 23,000+ genes, we can make many more proteins (the estimate is 2 million!). We all get a different arrangement of DNA coding for proteins, which makes us all different (other than identical twins, who have exactly the same DNA). For example, my DNA codes for proteins that determine my blue eyes, whereas my brother got DNA that codes for proteins that determine green eyes. We'll learn more about heredity when we get to genetics, but you need to know what nucleic acids are in order to understand a little about how they work.

ATP (adenosine triphosphate) is the third type of nucleic acid, and one of the most important molecules in the human body (second to DNA). Without ATP we would not be able to live. It is a monomer nucleotide that transfers energy in cells. Energy is needed to power almost all of our biological processes. It is composed of a 5-carbon sugar (ribose) + the nitrogenous base adenine (ribose+adenine = adenosine) + 3 phosphate groups. ATP releases energy when the bond between two phosphate groups is broken (by hydrolysis) to make either **adenosine diphosphate (ADP)** + an unattached inorganic phosphate (Pi), or **adenosine monophosphate (AMP)** + diphosphate (PPi). The free phosphates that are taken off the ATP molecules in order to release energy are availabe to be reattached to ADP or AMP to restore the ATP molecule. The reactions are reversible, for example:

$$ADP + Pi + energy\ consumed \rightarrow H_2O + ATP$$
$$ATP + H_2O \rightarrow ADP + Pi + energy\ released$$

We are constantly adding a phosphate to ADP molecules to make ATP during cellular respiration in the mitochondria of all of our cells. (Some cells contain about a billion ATP molecules!) Breaking down one glucose monomer will yield roughly 30 ATP molecules if we have oxygen for the reaction.

$$Cellular\ respiration:\ O_2 + Glucose \rightarrow CO_2 + H_2O + 30\ ATP$$

If we have a billion ATP in some of our cells (less in others), and we have 80–120 trillion cells in our body, it boggles my mind to think of how quickly we are generating ATP necessary to survive! (Muscle cells in particular need vast amounts of ATP!) (If there is no oxygen available, we switch to anaerobic respiration, which yields roughly 5 times less ATP and forms lactic acid as a byproduct. We will talk about this further in the muscle chapter.) We've already discussed that when we consume more sugar than we need for cellular respiration, we convert it to glycogen (stored in muscles and liver cells) or triglycerides (stored in adipose tissue) for later use.

Plants make ATP from cellular respiration as well as during photosynthesis. Plants make ATP in the first stage of the photosynthesis (dark reaction) in which CO_2 is incorporated into sugar. They make more ATP during cellular respiration for their other metabolic processes. Many plants make more sugar than they need for cellular respiration as well, so they store the sugar in roots, tubers, and fruits (which we enjoy since they are so sweet!).

Review of the Four Organic Macromolecules (hydrocarbon polymers linked together by dehydration synthesis and broken apart by hydrolysis):

Proteins: Made from amino acids. Examples: enzymes, structural, transport, regulatory, defense, etc.

Carbohydrates: Building blocks are sugars. Needed for cellular respiration to make energy and for structure and storage. Examples: glucose, starch, glycogen, cellulose, etc.

Lipids: Most have fatty acids + glycerol. Needed for insulation, cushioning, energy storage, hormones, cell membranes. Examples: fats, phospholipids, steroids, waxes.

Nucleic Acids: Made from nucleotides, (DNA, RNA, and ATP).

DNA programs all of the cell activity by encoding for proteins, RNA takes DNA's message outside of the nucleus, and the messenger RNA code is translated into proteins by ribosomes. DNA to RNA to Protein is the central dogma of molecular biology.

ATP is the energy-storage molecule for cellular processes.

Chapter 3

Cell Form and Function

R obert Hooke first discovered cells in 1665 when he was looking at cork cells through a compound microscope. Many scientists followed his lead to examine, name, and discover forms and functions of cells and describe the relationship between cells and living things. Many scientists such as Hooke, Leeuwenhoek, Schleiden, Schwann, Virchow, and Dutrochet contributed to the cell theory.

Cell Theory: **1.** Every organism is composed of one or more cells.

                                 2. A cell is the smallest unit that has properties of life.

                                 3. All cells come from preexisting cells.

All cells have three things in common:

1. A **plasma membrane** encloses the cell.
2. Every cell contains **DNA** (free floating in cytoplasm for prokaryotes or enclosed in the nucleus for eukaryotes). DNA is the hereditary instructions of every cell.
3. Every cell contains **cytoplasm**, which is the material inside the plasma membrane, consisting of thick jelly-like fluid called **cytosol**, which has dissolved substances like glucose, CO_2, O_2, etc., important for metabolic activity. (This fluid is between the nucleus and the cell membrane in eukaryotes.)

After those three similarities, the cells break off into whether they are the **prokaryotic** ("before nucleus"), very simple cells, older on the evolutionary time scale; or **eukaryotic** ("true nucleus"), more complex and advanced cells, evolved 2 billion years after the prokaryotes. Here are some basic characteristics of prokaryotic versus eukaryotic cells.

Prokaryotes	Eukaryotes
Evolved 3.5 bya	Evolved 1.5 bya
Lack a nucleus, free DNA	DNA enclosed in nucleus
No membrane-bound organelles	Membrane-bound organelles
Free ribosomes	Free ribosomes or on endoplasmic reticulum
Aerobic or anaerobic metabolism	Primarily aerobic metabolism
Some have flagella and cell wall	some with flagella/cilia and cell wall
Examples: Archaea and Bacteria	Plants, Animals, Fungi, Protista

Looking at a variety of prokaryotic cells, we see differences in the cell wall or membrane composition (more lipid versus more carbohydrate content), or differences in the types of proteins that are made, whether the cells can photosynthesize or not, whether or not they have flagella, etc. There are even more differences looking at a variety of eukaryotic cells. They have different shapes, sizes, functions, etc., etc. Most cells are microscopic. For example, you could line up 2,000 blood cells across your thumbnail! Cells are generally extremely small to maximize surface-area-to-volume ratio, which is important for regulating the internal cell environment. There are, however, extremely large cells as well; the longest being nerve cells (some can stretch over 3 feet long in humans and over 10 meters long in giant squid!). Our nerve cells have tiny helper glial cells to maintain metabolic functions so they can afford to be larger in order to conduct signals throughout the body. An adult human body is made up of roughly 80 to 120 trillion cells! There are over 200 different types of cells. Some general examples of cell types are blood (red and many types of white), liver, kidney, nervous, muscle, bone, epithelial, gland, endocrine, and fat. For now, we will just look at the structure and function of different parts of a typical animal (eukaryotic) cell.

Some organisms have a **cell wall** (outside the cell membrane), usually composed of a carbohydrate (cellulose for plants, or chitin for fungi), or carbohydrate derivative (peptidoglycan for bacteria). Animals, however, do <u>not</u> have a cell wall, so we will start with the cell membrane.

Cell membranes (also called the plasma membrane) are made from phospholipids that have two **hydrophobic** (water-fearing) fatty-acid tails and a **hydrophilic** (water-loving) head. When phospholipids are put into any fluid with water, they arrange themselves to have all heads toward the fluid, and all tails are toward each other in two layers, the **phospholipid bilayer**. (Fluid on each side of cell membranes always contains water.) Hydrophobic interactions as well as other non-covalent bonds hold the phospholipid bilayer together.

The inside of the cell is called the **cytoplasm.** The cytoplasm refers to the space from the cell membrane to the nuclear membrane and consists of thick jelly-like fluid called **cytosol** along with all of the cells' organelles. The cytoplasm is composed mostly of water. It also contains proteins, vitamins, minerals, nucleic acids, amino acids, ions, sugars, and lipids. Most **organelles** are membrane-bound compartments. The only organelles that do not have cell membranes are ribosomes and centrioles. Organelles are the "little organs" of the cell and carry out tasks to keep the cell alive and maintain homeostasis. These specialized units contain enzymes (proteins) that help run the reactions of the organelles. If the enzymes were not contained in their compartments, some of the different functions would conflict with one another—some enzymes break down molecules (hydrolysis), others build them (dehydration synthesis)—and the cell would not function properly. For example, if the enzymes from the digestive lysosome organelles were free in the cytoplasm, they would digest the cell from the inside out. Basically, the overall role of organelles is to **isolate chemical reactions** by providing separate locations for activities that occur within the cell.

Here is a depiction of a cell with all of the main organelles:

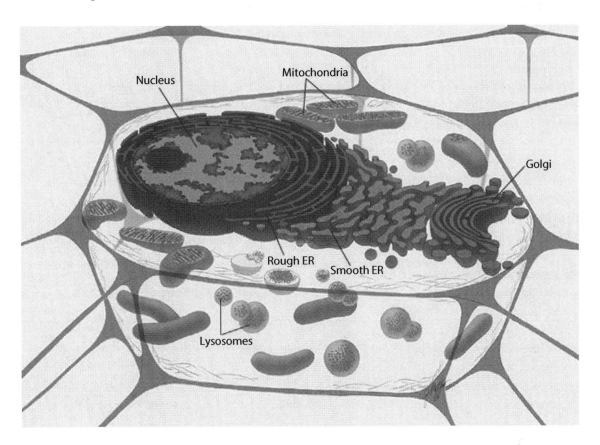

Nucleus is the "mastermind" of the cell since it houses the genetic material (DNA), which codes for RNA within the nucleus, which then takes the message out of the nucleus to code for all of the cell's proteins, which in turn determine the cell's functions. It is surrounded by the **nuclear envelope**, which is made from two lipid bilayers embedded with proteins and pores to let some molecules (like proteins and RNA) into or out of the nucleus. The DNA, however, cannot penetrate the nuclear envelope and therefore never leaves the nucleus. The DNA of each somatic cell consists of 46 chromosomes that would span six feet if we could stretch the chromatin (uncoiled form) end to end. (When they condense/thicken during cell division, they are called chromosomes and look like what you see with a karyotype—discrete particles within the nucleus.) The chromatin/chromosomes actually contain both DNA and proteins, and they are contained in the nucleoplasm of the nucleus. There is another area of the nucleus called the **nucleolus**, which makes ribosomes from ribosomal RNA (coded from DNA) and proteins (made in the cytoplasm). Two ribosome subunits are made in the nucleolus and then they are transported to the cytoplasm. Once they are assembled to form a functioning ribosome, they cannot re-enter the nucleus.

Centrioles are usually found right outside of the nuclear envelope and make the microtubules/filaments that are necessary for the cytoskeleton as well as for cell division. The centrioles also act as an anchor for the cytoskeleton.

Cytoskeleton is made up of microtubules, microfilaments, and intermediate filaments to give the cell its shape and structure (think of it as the backbone of the cell). Microtubules sometimes help move the cell as well if the cell has cilia (short tubules) or flagella (long tubules).

Mitochondria are the "powerhouse" of the cell and generate <u>ATP</u> (energy) through cellular respiration (remember using O_2 + glucose to make energy, water, and CO_2). Some cells have hundreds to thousands of mitochondria to ensure the production of enough energy for the cell. Mitochondria are self-replicating and contain their own mitochondrial DNA.

Ribosomes are the manufacturing plants that synthesize and modify all of the body's proteins based on the instructions from the DNA (delivered to the ribosomes by the messenger RNA). Ribosomes lack cell membranes and can be "free," in which case they are floating unattached in the cytoplasm, or they can be attached to the rough endoplasmic reticulum. Ribosome subunits are made in the nucleolus and then brought together by transfer RNA in the cytoplasm.

Rough Endoplasmic Reticulum is attached to the nuclear envelope and helps ribosomes make proteins. The rough ER can also modify proteins and send them out to other parts of the cell. There is another kind of Endoplasmic Reticulum (ER) …

Smooth Endoplasmic Reticulum makes lipids (like cell membranes), degrades fats, and inactivates toxins. The smooth ER does <u>not</u> have ribosomes associated with it.

Golgi Complex is like the UPS of the cell, since it modifies, sorts, and ships proteins and lipids for insertion into other organelles or out of the cell. (Material is put into various membrane-bound organelles called **vesicles** and **vacuoles** for storage, transport, or disposal.)

Lysosomes are specialized vesicles that bud off the Golgi to break down and recycle materials. They contain hydrolytic enzymes (which require a low pH environment of about 5) that the cell uses to digest macromolecules. The organic monomers are then returned to the cell for reuse. Lysosomes can engulf whole organelles if necessary and can also digest bacterial invaders.

Peroxisomes are another type of vesicle from the Golgi complex, which break down fatty acids, amino acids, and toxins. For example, peroxidase enzymes convert hydrogen peroxide (H_2O_2) to H_2O and O_2. Liver cells specialize in helping rid the body of toxins such as alcohol (ethanol) by making enzymes in **peroxisomes**, which convert toxic ethanol to less-harmful acetaldehyde and then acetate and finally carbon dioxide and water. It takes enzymes in peroxisomes approximately two hours to detoxify one alcoholic beverage. The detoxification process uses up oxygen, which means there is less oxygen in the cells for normal cellular activities. (Severe alcohol abuse causes liver cells to break down over time, which can result in liver failure and death.)

Differences between animal and plant cells include the fact that animal cells do not have a **cell wall**, or **chloroplasts** (which are the very important organelles in plants that contain the pigment chlorophyll for photosynthesis, and where the most important part of photosynthesis occurs). Plant cells can also have very large central **water vacuoles**. Animal cells can have vacuoles too, but animals generally have the smaller version called **vesicles**, which are mostly for storage and transport of nutrients and enzymes within the cell, and moving substances in and out of the cells through endo- and exocytosis (described in the next section).

Now let's talk about how substances cross into and out of the cell. The plasma membrane consists of a mixture of phospholipids and proteins, which have many functions (as receptors and channels, and they regulate transport, run reactions, etc.). In this way the membrane can control what enters and exits the cell and in what concentrations. Membranes are **selectively permeable**, which means they can let some things through, usually small, non-polar molecules, and keep other things out, like large molecules and ions.

Four ways small substances can cross the cell membrane through **passive transport**:
1. **Simple Diffusion**: Some substances can **diffuse** across the plasma membrane by passive transport (which requires no energy), when molecules or ions move down their concentration gradient (from high to low concentration). Some electrolytes and small sugars can cross cell membranes through simple diffusion.
2. **Osmosis**: Diffusion of water across a selectively permeable membrane. Water moves down its concentration gradient from **hypotonic** (more water) to **hypertonic** (less water). If too much water enters the cell, it will pop. This doesn't happen in our bodies because cells can regulate the solute concentration in the cell and therefore how much water will diffuse into or out of the cell. If there are too many solutes present in a cell, the cell will move solutes out. As the solutes leave, so does the water.
3. **Facilitated Diffusion**: Also passive, but involves transport proteins. These picky proteins move molecules/ions through <u>channels</u> into or out of the cell. For example, we have carrier proteins to move fructose into and out of cells.
4. **Bulk Flow**: Movement of fluid across a membrane can also be affected by the amount of pressure and solutes. Low solutes/high pressure results in high water potential; water generally will move from high water potential to low water potential (low pressure/higher solutes).

Three ways the cell can move substances across the membrane through **active transport** (requires energy (ATP)):
1. **Membrane Pumps** use the enzyme permease to move substances, usually in the opposite direction of diffusion (against the concentration gradient).
2. For **larger substances** to cross a cell membrane, the cells use membrane-bound organelles called **vesicles** (sometimes called vacuoles), which are formed through **endocytosis** (membrane pinches vesicle toward the inside of the cell with materials inside the vesicle). If the vesicle brings in solid organic material it is called **phagocytosis** (cell-eating) versus liquid material, which is called pinocytosis (cell-drinking). Fat is generally absorbed into cells by phagocytosis.
3. **Exocytosis** is the opposite; vesicles are pushed towards the outside of the cell, in order to move material out of the cell.

Immune System

Okay, now we're going to start our march through the different systems in the body. There are eleven organ systems that we will cover: immune, integument, muscular, skeletal, circulatory, respiratory, digestive, excretory, nervous, endocrine, and reproductive. You will notice along the way that I bring up topics more than once and will remind you of things you learned in other chapters. This is because many of these systems are so interconnected that I can't just talk about one system alone. You will see how the circulatory system, for example, is closely linked with every system of our body and how the nervous and endocrine systems control all the other systems with hormones, nerves, and neurotransmitters. Another system that deals with the entire body is the immune system, so we are going to start here.

Think of your body as having two very important highways running through it. One of them is the circulatory system: picking up and delivering nutrients, hormones, electrolytes, and oxygen, and purifying the body by helping to get rid of wastes such as carbon dioxide and toxins. The main cells in this highway are the red blood cells. There are other blood cells in our bodies, however, that are traveling in a separate system (the **immune system**). This highway is delivering white blood cells and proteins needed to fight infections all over your body, and includes lymph glands, where cells are stored, and stem cells in bone marrow, where cells are made.

We all have similar components to our immune systems but we build up immunity to different pathogens depending on which ones we've encountered and therefore which antibodies we make. Our immune systems have different strengths as well and this has to do with heredity and lifestyles. Some people are more resistant to certain invaders than others due to the genes they inherit. For example, recent research has shown that there is one form of a gene that actually provides higher resistance to HIV in some individuals. In terms of lifestyle factors, stress, smoking, not enough sleep, drugs, excess alcohol, and an unhealthy diet can all lead to a weakened immune system. Immunity decreases as we age, as well. In many species, choosing a partner to reproduce with has a lot to do with their immunity fitness. In order to have healthy offspring, individuals will often choose a mate who has a strong immune system. It is not a coincidence that many features that attract mates also coincide with immunity fitness in the animal kingdom. Birds are a

...ect example. Peacocks with the most eyespots generally have stronger immune systems than less-showy ...eacocks and they tend to get the most mates! In humans, we might not even realize that we are selecting a mate based on immunity but there are slight clues we can follow as well. Some research has documented how we can gather information about a person's immunity through pheromones (odorless chemicals emitted in sweat). This may be one of the causes of "love at first sight." Maybe this phenomenon is due to finding very compatible immune systems! This is all very fascinating to me, but for now I will move on to explain the immune system.

Let's start with what pathogens the body needs to protect itself against, and then move on to the different components of the immune system: main organs, glands, proteins, and cells and how they work to fight off infections. Finally we will discuss some immune system disorders, common diseases that threaten our society, and what you can do to protect yourself.

A **pathogen** is something that causes an infectious disease. Examples are **bacteria** (such as staphylococci, which cause staph infections; streptococci, which cause strep infections; and Escherichia coli); **fungi** (athlete's foot, ringworm, and yeast infections); **protista** (can cause diseases such as malaria and giardiasis); **animals** can either be vectors carrying a pathogen that causes a disease (such as mosquitoes (malaria), tsetse flies (sleeping sickness), and deer ticks (lyme disease)), or they can cause the disease themselves (roundworms can cause trichinosis); and **viruses** (cause many diseases such as the common cold, flu, HIV, mumps, measles, and chicken pox). Humans travel and carry infectious diseases with them everywhere. Viruses and bacteria can also travel via mammals, arthropods, birds, agriculture, water, waste, etc.

Bacteria are thought to be the most common type of pathogen (followed by fungi and then viruses), but I also want to make a note here that we have trillions of harmless as well as useful bacteria in and on our bodies as well, especially in our digestive system and on our skin. (Bacterial cells outnumber our own body cells 10 to 1!) Thousands of bacteria cover every square inch of our skin and help provide a barrier against more invasive bacteria. Other bacterial functions include helping with digestion, producing folic acid for us in our colon (which is a nutrient we can't make ourselves), and keeping other bacteria and fungi under control in places like the urinary system. Pathogenic bacteria typically represent a very small minority of the bacteria on our planet. (There are probably millions of species of bacteria on our planet, and fewer than 0.5% of them are pathogenic.) In fact, most bacteria are actually very beneficial to life on Earth in general. We already covered some basics about bacteria in the first chapter. Here's a review of some of the different types so that you'll remember they are not all bad.

Bacteria are generally divided into two groups based on their oxygen and carbon requirements. Autotrophic bacteria, such as cyanobacteria (blue-green algae), make their own food through photosynthesis and in the process supply a lot of oxygen for all living organisms (so they don't require oxygen). These bacteria obtain carbon from CO_2 in the process. The second type of bacteria is more prevalent and is called heterotrophic bacteria. They need to rely on others for food, and must obtain oxygen since they can't make their own. Many are decomposers and absorb nutrients from dead organisms and organic waste, and many fix nitrogen. They have a critical role of recycling nutrients to be reused (especially by plants). And then, of course, there are the heterotrophic bacteria that are pathogenic. Heterotrophic bacteria can be identified and classified by looking at the shape of the cells (coccus is round, bacillus is rod, or spirillum is spiral), and by staining the cells to determine what kind of cell wall and/or membrane composition they have (whether more lipid or peptidoglycan). This process is called gram staining. If the bacteria stains purple, it is called gram positive; if it stains red, it is gram negative. Gram negative bacteria are generally more harmful to humans because they have a thick lipid bilayer, which makes it harder for our body's immune system to destroy them. Our natural defenders against bacteria are antibodies, complement proteins, phagocytes, and T-cells. (We will cover how all of these work later in the chapter.) Gram negative bacteria are also less susceptible to antibiotics.

Antibiotics are drugs produced to target and kill bacterial infections. They work by interfering with cellular processes of the bacterial cells by either preventing protein synthesis, which blocks the bacterial DNA and RNA synthesis, or by breaking down cell walls or preventing new cell walls from forming. Meanwhile, evolution is occurring quickly in these prolific species and some bacterial strains are becoming resistant to our antibiotics. We are constantly updating these "miracle drugs" to keep up with the changes in the bacterial species. One thing that YOU can do to help stop the evolution of antibiotic-resistant bacteria is to **avoid using antibacterial soap**. Most of the bacteria we come into contact with are harmless to us and can easily be washed off with regular soap and water. Antibacterial soap is helping to create more virulent bacteria, since the .01% that are not killed by the soap can often mutate to resist the chemicals and propagate so that we will have more-resistant bacteria in the population. The soap is also contaminating our water supply and chemicals from it are ending up in fish and even breast milk! Another thing **not to put down the drain** is unused or expired antibiotics—you shouldn't have any anyway, since the next rule of thumb is to **always take the full prescription of antibiotics** (usually ten days). This ensures that all the bacteria causing the infection are killed, and leaves less of a chance that some remaining bacteria will develop antibiotic resistance.

Viruses are <u>not</u> cells and don't have all of the characteristics of a living organism, so they are hard to classify. They don't have a nucleus or organelles, but they do have nucleic acid—either DNA or RNA genome—plus a protective protein coat called a **capsid**. Many have receptor proteins on their surface to help attach themselves to their host cells where the virus will then inject its DNA or RNA in order to start the viral replication process. Sometimes the viral receptor proteins are so similar to our own cell receptors that we don't recognize them as foreign, which makes it hard (and sometimes impossible) to fight off the viral infection.

- They do not eat or respire, but they do reproduce with the help of their host cells—they are **intracellular parasites** because they rely on their host cell for metabolic and biosynthetic processes. A virus can make thousands of copies of itself and then the host cell will rupture and the new viral copies will go out, travel in the blood, and invade more cells. A virus must enter a cell to survive and reproduce.
- They can act on the host immediately or can remain dormant for years before taking over the host cells to create more viral particles. (HIV can remain dormant in helper T-cells for 10 to 15 years versus the common cold or flu, which have very short latency periods and cause symptoms very quickly.)
- Some viruses can alternate between dormant and active phases. For example, herpes symptoms will last several weeks and then subside, but the active phase can be triggered again by stress, diet, and sunlight.
- There are twenty-one families of viruses; classification is based on size, shape, chemical composition, whether they have DNA or RNA, and mode of replication. Some common examples: Ebola, herpes, smallpox, chicken pox, influenza, polio, HIV, shingles, rabies, and hantavirus.
- We generally think of viruses as very harmful since they can be the cause of cancer (by inserting oncogenes into the host DNA) or other serious diseases such as AIDS, but there are some viruses that

do not use human cells as their hosts but rather attack bacterial cells within our body. These are called bacteriophages. (These can be harmful or beneficial depending on which bacteria they attack.)

- Our immune system antibodies and T-cells help to destroy viruses and cells infected with viruses. Vaccines will help to create antibodies and white blood cells to fight future viral infections. A fever will slow the viral invasion down since viruses don't reproduce well at higher body temperatures. Antibiotics don't help to combat viral infections at all and should be taken only for bacterial infections. (Antibiotics cannot help with viral infections because viruses do not have cell walls, metabolic pathways, or protein synthesis to disrupt.)

There are two other infectious pathogens similar to viruses. Viroids contain RNA but no capsid, and cause plant diseases; and prions have protein with no DNA or RNA, but can still cause disease by affecting the central nervous system (ex: cause mad cow disease).

Vaccinations are incredibly important for preventing the spread of both bacterial and viral diseases. Of all of the medical advances in the last century, vaccines have probably saved the most lives. The U.S. now has over 17 vaccinations available and is 80–100% free from many preventable diseases such as small pox, polio, diphtheria, mumps, hepatitis A & B, tetanus, whooping cough, and chicken pox. A relatively new vaccine targets HPV and is very important to get in order to reduce the risk of cancer. Vaccinations have been blamed for autism, attention deficit disorder, brain damage, and multiple sclerosis; however, the vast majority of these cases have been unfounded. It is widely accepted that the benefits of vaccinations far outweigh the risks.

A **vaccine** is a very mild, inactive, or synthetic dose of a bacterium or virus. When the vaccine is delivered into the body's system (through injection, nose spray, or drops), the body's immune system will wipe it out as well as build up a memory for the pathogen by making both memory B cells and antibodies specifically for this pathogen. If the body encounters the pathogen again, the memory B cells will trigger the production of more antibodies to either destroy the invader themselves or flag T-cell to kill the pathogen. The immune system wipes out the pathogen usually before there are even any symptoms. Most vaccinations that we get as babies and children will last for many years, but antibodies and memory B cells don't last forever, so it's good to get booster shots to protect yourself. We especially need boosters for tetanus, HPV, diphtheria, whooping cough, pneumonia, shingles, and hepatitis A & B.

We can also get a vaccine to help prevent the flu every year. (It can take up to 2 weeks to build up the necessary antibody protection needed to fight off the influenza virus, so it's good to get the shot in October, before the flu season is in full swing. The vaccine lasts about 6 months.) We don't get a vaccination for colds because there are so many varieties out there and they change every year. The flu, however, is more severe, and scientists can pinpoint the most prevalent strains of influenza in a community a little more efficiently. Sometimes people will still get the flu after receiving the vaccine. This happens if they become infected with a strain of the virus different from the vaccine. (Remember, vaccines are extremely specific, and they will not cover all strains of a particular pathogen.) It is good to get a flu shot every year, especially if you have any upper respiratory ailments or a weakened immune system. Children should get vaccinated because their immune systems are not fully developed yet, and the elderly should get the flu shot because the immune system weakens with age. (Eggs are used to make the influenza vaccine, so people with egg allergies should not get the flu shot.)

Okay, now that we know who the invaders are, let's see how our body fights them off. We have an amazing response system to detect and destroy a variety of pathogens on a number of different levels (barrier, non-specific, and specific immune responses). Since pathogens are everywhere, we can't really avoid them, but we have multiple defenses against them. Our immune system is highly organized and consists of organs, lymph tissue, glands, and different types of white blood cells and proteins, many of which carry out highly specific functions. Our immune system is the body's overall ability to resist and combat pathogens and anything else that is non-self. The main cells that fight these invaders are different kinds of white blood cells. They are made in bone marrow from stem cells. They are then transported around the body, along with defense proteins, in the lymphatic system. The lymphatic system carries lymph tissue (consisting of H_2O, dissolved substances, proteins, and white blood cells) in lymph vessels all throughout the body.

We can recognize that something is non-self and needs to be destroyed based on the particles that are on the surface of the invader. These particles are called **antigens** and they are usually proteins, lipids, or oligosaccharides. The term is from "**anti**body **gen**erator," because antigens prompt the generation of **antibodies** and can cause an immune system response. We have self-antigens attached to all of our cells that we recognize. These don't cause an immune system response; the immune system attacks only non-self antigens. Foreign antigens are attached to **major histocompatibility complex** (MHC) proteins in cell membranes. There are roughly twenty-five families of MHC markers. T-cells learn to detect our own specific MHC markers in the thymus, which is very important because our T- and B-cells will attack anything with foreign MHC markers. (For organ transplants to be successful, the donor's and recipient's MHC markers on cells must be at least 80% the same, otherwise the organ will be attacked by the recipient's defense cells.)

Foreign antigens trigger an immune response because they are different from the particles we have on our own cells, self-antigens, which enable us to recognize our cells. For example, if you are type-A blood, you produce type-A antigens—specific proteins on the surface of your blood cells to help identify your cells as self so that your white blood cells will not try to destroy them. However if type-A blood is given to a type-B person, the introduced blood will be targeted and destroyed since someone with type-B blood will recognize A-antigens as foreign (a type-B blood person makes B-antigens and type-A antibodies). **Antibodies** are proteins found in blood or other bodily fluids (lymph) of vertebrates and are used by the immune system to identify and neutralize foreign invaders. Antibodies are made by plasma cells (derived from B-cells), in the spleen, one of the primary immune system organs. Antibodies bind to non-self antigens to flag pathogens or cells for destruction by white blood cells. Some antibodies can attack invaders directly, as well. Antibodies bind to the antigens on the outside of the pathogens or our own cells that contain pathogens inside (antibodies can't go inside cells). I know this is confusing at first, but we will go through all of the parts of the immune system separately and then I hope it will become clear how the parts fit together.

Immune System Tissues, Organs, and Functions

The lymphatic system has three main functions: defense, delivery, and disposal. It is composed of **lymphatic tissues** and **lymphatic organs**. The lymph tissue moves around the body in **lymph vessels**. It is composed of immune system cells (many types of white blood cells [**leukocytes**] made by red bone marrow), and **interstitial fluid** (some of which was previously in capillaries). The interstitial fluid contains water, defense proteins, and other organic molecules.

The Main Functions of Lymph:

1. Deliver white blood cells to the circulatory system as well as to lymph nodes and other tissues involved in immunity responses.
2. Pick up water and solutes that leaked from capillary beds into the fluid between the cells (interstitial fluid) and deliver these substances back to the circulatory system.
3. Deliver fats absorbed from the small intestines to the circulatory system. (The circulatory system then delivers it to where it needs to go to be used for energy, for biosynthesis, or for storage as adipose tissue for later use.)
4. Transport cellular debris and foreign material from body tissues to **lymph nodes** for disposal.

The lymph vessels have smooth muscles and valves, just like the circulatory vessels, in order to keep the lymph moving in the right direction. The lymph travels through lymph nodes, picking up white blood cells and antibodies when necessary or dropping off debris, and then continues to the lower neck, where lymph fluid can enter the circulatory system by draining into veins in this region.

The Main Functions of the Immune System Organs:

Lymph nodes—are small bean-shaped organs located throughout the body and connected by lymph vessels. Clusters of lymph nodes are located in the neck, armpits, abdomen, and groin areas. They are disposal sites and storage sites. White blood cells accumulate in lymph nodes after being produced in the bone marrow, and debris from infected areas is brought to the nodes to be broken down for disposal. During infection, lymph nodes are enlarged with lymphocytes, macrophages, and pathogen debris.

Red bone marrow—the site of our very important **stem cells**, which produce all of our different types of cells, including the immune system's white blood cells such as lymphocytes and phagocytes.

Spleen—the site of antibody production (by plasma cells), disposal of old red blood cells and foreign debris, and the site of red blood cell formation in embryos. The spleen can be removed (splenectomy) if it becomes too enlarged, often due to the body fighting off a serious illness.

Thymus gland—many types of T-cells multiply, mature, and are stored here after being made in the bone marrow from stem cells. Other immune system organs and glands can be removed if they become enlarged or infected, such as the appendix, tonsils, and spleen, but we generally can't live without the thymus due to its important T-cell functions.

Tonsils—immune tissue at the back of the oral cavity that helps to fight off microbes from inhaled or ingested particles. If the tonsils become too enlarged from infections, they can be surgically removed (tonsillectomy).

Mucus-Associated Lymphatic Tissue (MALT)—located in the small and large intestines and the appendix. Half of the body's lymphocytes and macrophages are in this tissue to fight off pathogens in the digestive tract. (If there are too many pathogens for the MALT tissue to handle, the body will resort to flushing methods to get them out, such as vomiting and diarrhea.)

When the **appendix** is infected or enlarged (caused by too many white blood cells [WBCs] in the gland due to an infection in the area), it usually needs to be removed before it inflames so much that it bursts, spreading the infection in the abdomen. If the appendix bursts, the bacteria can get into the blood, causing sepsis and death. The surgery to remove the appendix is called an appendectomy. Although the tonsils, spleen, and appendix all play important roles in the immune system, we can survive without them.

All of these organs are factories for production and storage of lymphatic tissue cells and proteins, but they also serve as filtering stations to cleanse the lymph tissue before it is returned to the bloodstream.

Leukocytes (white blood cells)—made in the red bone marrow by stem cells and then deployed all over the body. Some live only a few hours to a few days (as opposed to red blood cells, which live approximately 120 days), and some live for years. They need to be produced constantly. How many WBCs are produced at any given time can depend on a person's age, health, and activity levels. WBCs comprise only about 1% of blood in the circulatory system, but the WBC count will rise when the body is under attack from an infection. The main types we will go over here are the macrophages, lymphocytes (smaller and more specialized), monocytes (can convert to macrophages), and dendritic (antigen-presenting) and mast cells (for allergic responses). There are many more types of leukocytes, but we can't cover everything in this overview.

In order to explain all of the types of white blood cells and proteins involved in the immune system, we also need to go over the three types of immune responses:

Three Basic Defense Systems

1. **Barrier systems**, non-specific, innate (we're all born with this), simple defenses to keep out disease-causing invaders. This first line of defense includes barriers such as:
   - Skin—very tightly packed epidermal cells that prevent microbes from entering (we're also continually shedding our skin, which sloughs off bacteria as well). We also have specialized phagocytes in our skin tissue called Langerhans cells that will engulf bacteria within the skin.
   - Secretions, such as tears, mucus, saliva, and urine all help to neutralize microbes that enter the body. Lysozyme enzymes within some of the secretions (in saliva and mucus) also help destroy bacteria.
   - Hairs and cilia (such as in respiratory system) help trap microbes.
   - Processes like sneezing, coughing, diarrhea, and vomiting expel microbes.
   - Special chemical factors such as lysozyme, sebum, and high-acidity gastric juices and vaginal secretions also help to kill pathogens and act as antimicrobial barriers.
   - Bacteria, although considered a pathogen in some cases, can also help. We have plenty of bacteria, especially in the digestive system, that are actually helping our immune system by killing other microbes.

2. **Non-Specific Immune System**, also innate, is the second line of defense. This system involves:

- Special white blood cells called **phagocytes** (made by stem cells in bone marrow such as in the ends of your femur bones), which can engulf pathogens in order to destroy them. Phagocytes are large white blood cells (ex: **macrophages**) that can engulf entire pathogens circulating <u>outside</u> body cells (within tissue fluid). There are proteins (complement proteins) that help in this process as well by flagging the pathogens to be ingested by the phagocytes. Macrophages can engulf hundreds of bacteria at once—think of a Pac-Man gobbling ghosts! Phagocytes also engulf dead or dying cells, debris, and foreign material and bacteria. There are also smaller phagocytes called microphages that work mainly in the nervous system.

- **Complement proteins** are secreted by the phagocytes and attach to the pathogen, which flags it for ingestion by the phagocyte. Some complement proteins can also kill microbes directly. Complement proteins surround the invaders and can cause the osmotic pressure to change in a way so that water enters the bacterial cells to destroy them. Complement proteins are also helpful with inflammation and antibody responses.

- Antimicrobial proteins called **interferons** are also important for the second line of defense since they protect host cells from viral infections. Their action is intensified during fevers.

- A **fever** is when the body temperature is above 98.6° F, although most doctors consider a fever to be closer to 100.4° F since many people run a little warmer than 98.6 (some people run a little cooler than 98.6° F too). Chemicals, such as cytokines released by phagocytes, can trigger the brain to raise the temperature of the body. This, in turn, causes chemical reactions to occur more quickly, and the immune system starts to work faster. Some pathogens cannot function in higher temperatures, which helps to destroy them (for example, some viruses can't reproduce as quickly in higher temperatures). If the temperatures get too hot, however, it becomes dangerous for our own cells as well. (Proteins, such as enzymes, denature at higher temperatures and if the enzymes are not running chemical reactions properly, the cells can die). This is why when you are sick a low-grade fever can help you fight your infection, but if your temperature gets too hot (above 102° F), you definitely want to cool yourself down with fever-reducing medicine.

- Other types of white blood cells such as NK cells and mast cells are also part of the non-specific immune system, but I will describe these once we cover the different types of lymphocyte cells in the next section. These are all part of the non-specific immune response because it doesn't matter what type of antigen is on the outside of the pathogen—as long as it is a foreign antigen, it will be attacked.

3. **Specific immune system** is the third line of defense and is highly specialized, since it targets specific pathogens that the body builds up immunity against. It is an adaptive immunity, not innate, since it must be developed over time. When you come into contact with a particular pathogen, your body will remember the antigens on this pathogen, so the next time you encounter it, you fight it off more quickly than the first time. (This is where vaccinations are important.) The specialized white blood cells involved with this line of defense are called T- and B-cell lymphocytes (also made in the bone marrow). These lymphocytes are much smaller than the phagocytes previously mentioned and travel around your body in lymph fluid (along with phagocytes) or in the circulatory system. When they are inactive (not fighting a pathogen), they are stored in white pulp of the spleen or in the thymus and lymph nodes. When they detect a pathogen, they initiate cell replication and communicate with other immune system components and go to work. Each lymphocyte has receptors for specific antigens. The

immune response for the specific immune system can either be **cell-mediated** where the lymphocyte attacks a cell directly, or it can be **antibody-mediated** where antibodies are produced by **plasma cells** (derivatives of B-lymphocytes) to help flag antigens for the phagocytes or lymphocytes to recognize. Specific receptor structures on the antibodies or WBCs will allow them to interact with the specific pathogen that they are attacking. Immunity can be active, where the body acquires immunity through contact of the pathogen, either by experiencing the actual invaders or by getting an immunization for the pathogen. Another way to acquire immunity is passively, such as when antibodies are delivered from a mother to her baby through the blood (in utero) or via breast milk. (This is one of the many benefits of breastfeeding.) A mother generally comes into contact with the same pathogens as her infants and therefore makes the necessary antibodies to fight off the invaders. Although these acquired antibodies provide immediate protection at no cost to the infant, they provide protection only until they are used or broken down; after that the infant must make his or her own antibodies to fight off infections. The acquired antibodies do not get cloned or stimulate antibody production themselves; that comes naturally over time. Infants can also get antibody mixtures in gamma globulin shots to provide protection for 3 to 6 months, to fight off different pathogens. They can start to get specific vaccinations as infants as well.

Types of Lymphocytes (Note that not all lymphocytes are involved in specific immune defense; some are involved in non-specific responses, but I'm going to list them all in this section.) T- and B-cells are the only defensive cells with receptors for specific antigens, so they are part of the **adaptive immunity response system**. T-cells are also important because they can get to the pathogens that invade <u>inside</u> cells by killing the entire cell that carries the pathogen, whereas the phagocytes can attack pathogens only outside of cells.

1. **T-cells** make up 50% of the lymphocytes circulating in blood and lymph. They are made in bone marrow and mature and are stored in the thymus (in response to thymic hormones) until needed for defense. (They are called "T" cells because of their location in the thymus.) Cytotoxic "killer T" cells are very powerful white blood cells that circulate throughout the body and are important for cell-mediated immunity responses by targeting entire cells rather than just pathogens between cells. They touch-kill invading cells with foreign antigens on the surface. Antibodies attach to the foreign antigens to target the infected or foreign cell, so that the T-cells recognize which cells to destroy. It is called "touch-kill" since they don't have to engulf the cell like macrophages do; instead they release deadly chemicals, which create holes in the cell. This destroys the target cell because the cell's cytoplasm leaks out of the holes, along with organelles, DNA, and pathogen, and the whole cell is destroyed. After it kills the infected (or foreign) cell, it moves onto the next infected cell. Sometimes healthy cells are killed in the process too. The T-cells attack all of the cells in an infected area to ensure they wipe out the invading pathogen. Macrophages will come clean up the debris by engulfing the contents of the spilled cells after the T-cells have done their job.

T-cells become activated by the presence of **Antigen-Presenting Cells (APC)**. Macrophages and dendritic and B-cells can all become antigen-presenting cells. These cells can engulf the pathogen, break down the antigens on the outside of the pathogen with lysosomes inside the APCs, and then move pieces of these antigens to bind with MHC markers in the cell membranes, so that the antigens from the pathogen are then on the cell membrane with the MHC markers to flag an antibody to come attach to the antigen. This in turn will flag the T-cell to come touch-kill the APC. (T-cells need to have an antigen presented to them in order for them to work and they have specific receptors for specific antigens.) Cytotoxic T-cells divide when they are stimulated by cytokines released from helper T-cells.

2. **Helper T-cells** help to bind the self-MHC marker with the foreign antigen on the APC, which in turn activates T- and B-cell responses; by activating T-cells and antibody production by plasma cells, T-cells can also induce phagocytes to help with the invasion. Helper Ts mature in the thymus along with the other T-cells. Helper T-cells also produce the proteins' cytokines, which have <u>many</u> functions (see protein section below). Helper T-cells help launch both the antibody-mediated immunity outside the cell and the cell-mediated immunity of T-cells, killing cells with pathogens inside. (The HIV virus targets helper T-cells, which is why patients with HIV are immunocompromised. Without helper T-cells, T- and B-cell responses and cytokinin production are all diminished; between those three factors, most of the immune system's defenses are wiped out.)

3. **B-cells** are also made in bone marrow from stem cells, but rather than traveling to the thymus, they spend most of their time in lymph nodes and interstitial fluid. In the interstitial fluid they act as antigen-presenting cells and in the lymph nodes they become **plasma cells** and produce antibodies (defense proteins), which target specific antigens. **Antibodies** circulate in the body (in blood, lymph, and other fluids), and either kill pathogens themselves or flag pathogens for destruction by phagocytes, lymphocytes, or complement proteins. Antibodies bind to <u>antigens</u> on the pathogen or foreign or infected cell. (Antibodies can't go inside cells.) If a pathogen is inside a cell, the T-cells can touch-kill the entire cell containing the pathogen, which destroys the pathogen as well. B-cells are said to facilitate antibody-mediated immunity. Antibodies are generally produced within 3 to 5 days after the infection occurs and can remain in the body for several months.

When B- or T-cells divide, some become **effecter** cells that act to destroy <u>now</u> (active B-cells are also called plasma cells). The B-cells that are saved for later are called memory B-cells.

4. **Memory B-cells** are important lymphocytes that you save for the next time you are infected with the same pathogen. They can recognize foreign antigens that they have encountered before and produce the specific antibodies that will help with the attack. This helps speed up the response time to subsequent infections. (It takes less energy for the body to respond to a pathogen the second time around.) This is where **vaccinations** come in handy. If you are given a vaccine or booster shot, it has the look and feel of the pathogen, so that your body will produce B-cells to save for later, but you don't usually get sick from the shot. Memory B-cells can last for 10 to 20 years. (This is why for some vaccinations you need a booster shot, usually 10 years after you receive a vaccination. This ensures that you will create more memory B-cells for that particular pathogen because your supply may be depleted.) We can recognize and remember millions of different pathogens!

5. **NK (Natural Killer) cells** are different from other lymphocytes in the fact that they don't need a memory of the invader to recognize an enemy. They combat bacteria, viruses, and cancer cells all the same way. They inject poisonous substances—such as perforin—into the bacteria, viral-infected cell, or cancerous cell, which then kills the invader. (This is called "touch-kill" or "cell-to-cell combat.") Then macrophages come clean up the debris. They are not specific immunity defenders since they do not

target specific antigens. In fact, they will also kill cells that are lacking self-antigens. They also do not respond to immunizations and don't produce memory cells.

Other Leukocytes

Mast cells (similar to basophil cells) secrete histamines, which cause **inflammation**; since the histamines trigger arterioles to dilate, more blood flows through the vessels, increasing the action of phagocytes, which speeds up the healing process. (During an inflammatory response, capillaries tend to leak so that the WBCs can get to the site of injury.) This is a local response causing swelling, heat, and pain, versus an all-over response of **fever**, which develops when macrophages release the proteins called **cytokines** that stimulate the brain to release prostaglandin, which helps to raise body temperature. (Some pathogens can't survive and reproduce in the higher-temperature environment produced by a fever. The main way that fever helps during infection is that by raising the body temperature, it also helps to speed up defense cells and reactions that help fight and repair infected areas—molecules and cells move faster in warmer temperatures. A fever that becomes too high, however, is dangerous because enzymes denature at very high temperatures and then metabolic processes break down.) Fever and inflammation are both non-specific immune defenses like phagocytosis by the macrophages.

Dendritic cells alert the adaptive immune system when an antigen is present in tissue fluid, skin, or body lining. A dendritic cell can also become an **antigen-presenting cell (APC)**.

Monocytes have two functions. Under normal conditions they help replenish macrophages and dendritic cells to keep in reserve (half are in the spleen and half circulate in the blood and move into other tissues). Or they can help to fight an infection by moving quickly (within 8–12 hours) to the site of inflammation to divide and differentiate into macrophages or dendritic cells.

(There are more leukocytes, I'm just not going to go over all of them here.)

Defense Proteins (I've already mentioned most of these, but just to summarize them):
1. **Complement proteins**: Flag pathogen for phagocytes to engulf, or some complement proteins can kill pathogen directly. They can help with antibody responses and inflammation as well.
2. **Cytokines**: (made by helper T-cells or phagocytes) Have many forms and functions: used for cell–cell or cell–tissue communication to coordinate immune responses; cause inflammation, fever, and stimulate bone marrow stem cells; cause T- and B-cells to divide and specialize; attract phagocytes to pathogens; and activate NK lymphocytes to kill cancer cells and cells infected by viruses. Some cytokines can kill tumor cells by themselves, and can cause T-cells to accumulate in lymph nodes during an infection and stimulate the cell-mediated immune response. Examples: Interleukins, Interferons, and Tumor Necrosis Factor.
3. **Major Histocompatibility Complex (MHC) markers**: Found on the outside of all cells to help with self-recognition or foreign recognition (ex: when APCs attach the foreign antigen pieces to the MHC markers). No two people have the exact same MHC proteins (other than identical twins). MHC I are found on all cells except red blood cells and activate cytotoxic T-cells; MHC II are found only on APCs such as macrophages and dendritic and B-cells. These markers are activated by helper T-cells.

4. **Antigens**: Usually proteins or saccharides that help the immune system determine self versus non-self. Antigens can also be parts of the invading microorganism, such as the capsid, flagella, cell wall, toxins, or capsules. Foreign antigens trigger the production of antibodies, which can flag the invader for destruction by WBCs or the antibodies can neutralize the invader on their own.

5. **Antibodies** can kill invaders or attach to antigens to signal for white blood cells to kill pathogens. Antibodies can also turn on complement proteins. Antibodies are produced by plasma cells (which are derivatives of B white blood cells) in the spleen. They are very important for specific immunity responses. Antibodies have specific receptor sites for each individual antigen that has triggered its production. The antibody binds to the antigen with a lock-and-key fit. Vaccinations trigger the production of antibodies so that we will be able to combat diseases faster and more efficiently if we ever come into contact with the real pathogen.

Five classes of antibodies collectively called immunoglobulins:

1. **IgM** is important for newborns, since it is made the most out of the five types. It is also the first antibody released in any immune response.
2. **IgG** is 80% of antibodies in blood.
3. **IgA** is found in saliva and other exocrine gland secretions. When IgA is attached to mucus cells, bacteria and viruses can't attach to these cells, which helps with the barrier defense system. These antibodies can stop pathogens before they enter the bloodstream.
4. **IgE** helps with allergic response (triggers mast cells to release histamines).
5. **IgD** is found on B-cells and binds with antigens that help stimulate B-cell activation.

Immune System Disorders

1. Allergies: The immune system usually distinguishes between pathogens and other harmful chemicals in the environment, but sometimes a person's body will react to a harmless substance as if it were an antigen. Then the body produces antibodies to stimulate mast cells to release histamines or prostaglandins (chemicals that cause blood vessels to expand, the eyes to produce tears, the nasal passages to produce/secrete mucus). The harmless substance is referred to as an **allergen**; the response is called an **allergy**. Some people are genetically predisposed to allergies. Infections, emotional stress, and changes in air temperature can also cause allergic reactions. An extreme allergic reaction can cause circulatory collapse and suffocation as airways constrict to a point where you can't get enough oxygen to the lungs. This is called anaphylaxis and it can be lethal. (About 1,500 people a year die from anaphylaxis.) A person can go into **anaphylactic shock** after coming into contact with the allergen that they are hypersensitive to (often a particular food like nuts or eggs, some medications, venom from bee stings, etc.). The allergen triggers the release of inflammatory mediators from the mast cells. This can cause itching, swelling, hives, etc., but can also cause airways to close, fluid to gush from blood vessels, dramatic decrease in blood pressure, and can lead to the collapse of the cardiovascular system. Patients need immediate treatment such as oxygen and epinephrine. Epinephrine is a powerful antihistamine that will quickly open the airways and restore the blood pressure (it is usually injected in the thigh, because you don't want to inject it too close to your heart).

2. Autoimmune disease is when the immune system targets the proteins, cells, tissues, or organs of a person's own body. Many of these diseases are inherited. Abnormal genes are passed down, but sometimes people develop autoimmune diseases due to drugs, environmental factors, hormone problems, or viruses that alter the host DNA. Often the cause is unknown. There are some populations who are more at risk for or affected by certain autoimmune diseases than others. Females are generally more affected than males. Autoimmune diseases can target different areas of the body:

Rheumatoid arthritis (RA) tends to affect Native Americans the most and attacks tissues in joints. People with this disease are genetically predisposed to it. Their macrophages, T- and B-cells, become activated by self-antigens on joint cells. The immune system attacks the joint tissues and inflammation occurs. Eventually joints become immobile and bones can become misaligned or even fused together.

Lupus is more common in African American and Hispanic women (nine times more common in women than men) and affects many different parts of the body: skin, lungs, kidneys, and joints. The immune system attacks different tissues, causing inflammation and tissue damage.

Type 1 Diabetes (mellitus) is where the immune system T-cells attack and destroy insulin-secretory cells in the **pancreas** so the body doesn't secrete enough insulin. (Remember that insulin is a hormone that regulates body cells to absorb glucose from the blood in order to be used for cellular respiration to generate energy.) If the body is not absorbing glucose from the blood, it is lost with the waste (urine) and needs to be constantly replenished. People with diabetes need to check their glucose levels and administer insulin shots in order to help them absorb the proper amount of glucose from their blood.

Multiple sclerosis is where a person's immune system attacks the myelin sheaths around the axon of neurons. This severely damages the nervous tissue and causes symptoms such as problems with motor coordination, speech, vision, swallowing, etc. The cause of this could be environmental, genetic, or a viral infection. Research is still being done to try to pinpoint why this happens. More than 400,000 Americans have this disorder and there is still no cure. There is now, however, medication that can slow the progression of the disease.

Crohn's disease is an autoimmune disease where the immune system attacks part of the digestive tract, anywhere from the mouth to the anus, but usually in the small intestines. Nutrients are not properly absorbed from the digestive system to the circulatory system and it can cause weight loss, vomiting, inflammation, pain, diarrhea, and other symptoms. It is often genetic but can be environmentally induced as well. It usually doesn't appear until the teens or 20s, but can occur at any time. Men and women are equally likely to suffer from this disease; smokers are three times more likely to be affected than non-smokers.

Celiac disease affects 1 out of every 133 people in the United States. People with this disease are highly sensitive to gluten, which is a protein found in wheat, rye, and barley. Their immune system responds to gluten by attacking and damaging the small intestines, making it hard to absorb nutrients and causing discomfort in the digestive tract as well as in bones and joints. It can cause osteoporosis, arthritis, fatigue, anemia, numbness, anxiety, and depression. Eating a healthful gluten-free diet will dramatically improve these symptoms. There is more awareness of the prevalence and seriousness of this disease now, so there are more options in grocery stores and restaurants for gluten-free foods, but gluten is still a challenge to avoid.

3. **Immunodeficiency** can be acquired or inherited, and results in a weakened or non-functioning immune system. Some of the inherited causes are due to flaws in the immune system (such as lacking a thymus or problems with B-cells and antibody production). Some extreme cases have been documented where there is a lack of all of the immune system functions.

Cancer of the Immune System can come in different forms, such as **lymphoma** (cancer of the lymph glands) or **leukemia** (cancer of the blood or bone marrow). Both of these cause major problems with the proper functioning of the immune system in fighting off other infections as well. We will cover this subject more in Chapter 18.

AIDS (acquired immunodeficiency syndrome) is a type of immunodeficiency that is acquired through the **Human Immunodeficiency Virus** (HIV). HIV is a retrovirus, which means it has RNA rather than DNA within the protein capsid. The virus, identified in 1981 and thought to have started in Africa from a similar strain that infected chimpanzees and monkeys, inserts its RNA into the lymphocyte helper T-cells (called CD4 T-cells). These have a specific receptor on the surface that the HIV attaches to (HIV can also target dendritic cells and macrophages, but mostly helper Ts). Once the HIV's RNA is injected into the host cell, it goes through reverse transcription to make DNA. This DNA is then inserted into the host's chromosome, and can either remain dormant for years (10–15), or start using the cells right away as viral factories to create more HIV particles (the viral DNA codes for new RNA that goes into the cytoplasm to make viral particles). The new viral particles then bud off the host cell and are released into the bloodstream, ready to start destroying more white blood cells. The "viral load" is the number of viral particles in the blood. Once the number of viral particles exceeds the number of T-cells, the person will start suffering from chronic infections. A person is not said to have "AIDS" until the CD4 T-cell count is below $200/mm^3$ of blood.

When the helper T-cells are acting as viral factories, they can no longer perform their regular functions (helping to activate T- and B-cell responses), so the immune system breaks down. This leads to chronic infection, fatigue, nausea, weight loss, and enlarged lymph nodes. Death usually results from one of the typical AIDS-related diseases such as tuberculosis or cancer that the body is unable to fight off due to the weakened immune system caused by HIV. There is still no cure for HIV, so **prevention is critical**. HIV has infected ~33 million people worldwide since the 1980s and has resulted in 25 million deaths. There are about 1.2 million people infected with HIV in the United States and roughly 500,000 of them have AIDS.

HIV is transmitted when infected blood, semen, or vaginal secretions enter another person's bodily tissues. Common ways this occurs are through unprotected sex and shared needles of drug users. Contaminated medical equipment, razors, toothbrushes, ear piercing or acupuncture needles, and pedicure and manicure equipment can also transmit the virus, but only if they are used within twenty minutes of being infected; otherwise the virus will become inactive since it is unstable in the presence of oxygen. HIV can also be passed from mother to baby during labor (if infected blood gets into the baby's mouth or eyes) or from breastfeeding. HIV is not passed through food, air, or water. Infected adults are dying from AIDS more than any other single cause of death, especially youths (ages 15–24). Over 10 million youths are living with

AIDS around the world, mostly in developing countries where there is less health care education and less access to contraception such as condoms.

HIV cannot be eradicated by medicine, but there are "cocktails" of anti-HIV drugs to help slow down the HIV-replication process and suppress the virus (they are very expensive). Some drugs help to stop DNA from coding the HIV proteins and some help inhibit enzymes that make the viral proteins. HIV replicates and mutates so quickly that it is hard to make a vaccine for it as well, but there are 35 vaccines being tested around the world (none successful yet). Recent research has found a mutated gene in some people's DNA that makes them either immune or more resistant to HIV. People who have the mutated gene from both parents are immune to HIV but this only occurs in about 1% of people with northern European heritage (they think the gene mutation occurred in Scandinavia). If you possess only one of these mutated genes, you are more resistant to HIV but not immune (you might develop it more slowly than most people). This occurs in 10–15% of people with northern European heritage. I'm sure we will hear more about this in the future as more people map their genotypes to find out their susceptibility for different diseases and disorders. Currently the best method to stop HIV is **PREVENTION and EDUCATION**, so spread the word, educate your friends, and definitely practice SAFE SEX (the safest barrier method is using a condom. If you're trying to prevent pregnancy as well, then using oral contraception and a condom are the best combo).

I'm not going to go over all of the amazing varieties of infections that affect people worldwide, but I do want to mention some other infections for you to be aware of.

Common Sexually Transmitted Infections (STIs)

Chlamydia is the most common STI and is caused by bacteria. Three million Americans get chlamydia per year, the majority of whom are under the age of 25. It affects both women and men. Ninety million people around the world have this infection. Symptoms include burning when peeing, discharge from genitals or rectum, and/or tenderness or pain in genital area. Many people don't realize that they carry the infection since they don't have any symptoms. They often find out from their partners whom they've already infected, if their partners do show symptoms. Bacteria can migrate to the lymph nodes and cause inflammation there as well.

Pelvic Inflammatory Disease (PID) results when the bacterium enters the vagina instead of the urethra and affects the reproductive organs, which can lead to sterility. Some women need to get a hysterectomy (have their uterus taken out) if the infection has moved into their reproductive tract.

Gonorrhea, also known as "the clap," is another common STI caused by bacteria and can be treated with antibiotics. (There are 16 different strains of this bacterium.) Symptoms include pain in the genitals and during urination, and discharge. It can also lead to PID, sterility, and joint and valve problems if left untreated. (The name has a few origins but one of the weirder ones I've heard is that doctors would "clap" the discharge out of the penis with a heavy instrument—yikes!) More likely the name comes from the French word for brothel, which is "clapier."

<u>Syphilis</u> is another bacterial infection in which the bacterium causes skin ulcers; rashes; lesions in eyes, bones, nervous system, liver, and aorta; and can lead to paralysis, insanity and death. Some strains are antibiotic-resistant so syphilis can be very dangerous.

<u>Human Papilloma Virus (HPV)</u> causes genital warts and can cause cervical, anal, vaginal, penile, throat, and oral cancers! There are vaccinations now that guard against four of the most common strains of this virus, recommended for girls during puberty (before they become sexually active). Women should have their pelvic exam (PAP smear) done regularly to check for cervical cancer once they become sexually active, since it is the fifth leading cause of cancer-related deaths in women. Recent studies have noted that throat cancer, due to this virus, is on the rise in men, so there is a question about whether or not to start giving it to boys during puberty as well.

<u>Trichomoniasis</u> is caused by a protista and is the most curable of the sexually transmitted diseases (it is treated with antibiotics even though it is not due to a bacteria). There are 7.4 million new cases of trichomoniasis a year! Symptoms include painful urination or ejaculation, itching, or yellowish discharge. Some people never have any symptoms and therefore pass this along without realizing they have it.

<u>Pubic Lice</u> (also known as "crabs") are just what the name implies … lice in pubic hair. This causes itchiness, scabs, and general discomfort, and is treated with topical creams and shampoos. These insects can also be transmitted when people share beds, towels, and clothes; since the lice can jump from host to host, it's not transmitted just by sex.

<u>Genital Herpes</u> (type 2) is the second-most common type of herpes after type 1 oral herpes. Twenty percent of the U.S. population of adolescents and adults has one form or the other. Both types of herpes are caused by viruses. They are transmitted through saliva, genital fluids, or through contact with lesions. There is no cure. There are periods when the virus is dormant and no symptoms are detected, and then certain triggers (hormonal, stress, and weakened immune systems) can cause the symptoms to flare up. Oral sex can transfer type 1 herpes from the mouth to the genitals (and cause lesions there) or vice versa—the genital strain can be transferred to the mouth and cause cold sores.

There are more sexually transmitted infections but I will stop here … the bottom line is that practicing safe sex and limiting the number of partners will reduce your risks of contracting sexually transmitted infections.

Patterns of Infectious Disease

How Pathogens Spread:
1. **Direct Contact** with a pathogen, touching bodily fluids of infected partner (hands, mouth, genitals). ex: HIV, herpes
2. **Indirect Contact**: Doorknobs, tissue, food, water. ex: Giardia
3. **Inhalation** of viruses that are spewed into the air by uncovered coughs and sneezes. ex: common cold, flu, tuberculosis
4. **Contact with a vector**: Mosquitoes, flies, fleas, ticks. Vectors carry pathogens from infected persons (ex: plasmodium protist causes malaria)

Best ways to avoid getting sick and stop the spread of pathogens:

Wash your hands regularly (with soap) for at least 20 seconds. This will help especially with the common cold virus.

Get vaccinated to build up your immunity and stop the spread of diseases.

Contain cough/sneeze (use your upper arm rather than your hand to cover your mouth).

Practice safe sex (condoms for the best barrier protection).

Don't share needles (best not to do drugs, of course!).

Good sanitation.

Cook food thoroughly and properly.

Exercise reduces the risk of getting sick.

Eat a healthy diet with plenty of fruits and veggies (which have vitamins and minerals that help your immune system) to help to ward off infections.

Eat foods with probiotics (good bacteria) like yogurt. These bacteria can help your system ward off bad bacteria and fungi.

Be sure to get enough sleep, which helps your body with repairs and immune defenses.

Reduce stress (high levels of stress hormones can weaken the immune system).

Chapter 3

Tissues, Cavities, Membranes, and Skin

Remember in the first chapter when we looked at the hierarchical order of things? The smallest living unit of an organism is the **cell**. We've already gone over the different parts of the cell and the functions of the different organelles, but where do they come from? All cells in our body come from "stem cells." These are the first cells to form when the sperm fertilizes an egg and the zygote starts dividing. Embryonic stem cells differentiate to give rise to more than 200 types of cells, such as muscle, ligament, cartilage, neuron, red blood cell, and white blood cell (different types of phagocytes and lymphocytes). (Remember that the human body is composed of 80–120 trillion cells!) This is happening in the very first stages of the developing embryo when the dividing cells are differentiating into the three germ layers endoderm, mesoderm, and ectoderm, which give rise to all of the different types of cells in the body. Adult stem cells (found in bone marrow and umbilical cord blood) make cells that are constantly being regenerated, such as red and white blood cells, skin cells, and some types of digestive tissue cells. Even though they are called "adult" stem cells, these cells are also found in fetuses, children, and adults. It just means that they are not the cells present in the very beginning stages of embryos. There are also amniotic stem cells found in amniotic fluid, which can give rise to fat, muscle, bone, liver, and nervous and epithelial cell lines.

Stem cell research is an exciting new field of medicine, which will open up many doors for improving human health if it can be used for tissue repair and replacement. Some are hoping that it will help combat diseases such as cancer and some nervous tissue diseases such as Parkinson's and spinal cord injuries. Some adult stem cells can give rise to certain tissues like blood, cartilage, and muscle, but using embryonic stem cells is more effective. These cells are usually taken from one-week-old embryos from fertility clinics (aborted embryos from natural pregnancies or donated egg and sperm to generate embryos for this purpose), but it is very controversial research, since the embryo is destroyed in the process. Other research, however, is being conducted where single cells are being plucked out of two-day-old embryos and the embryos are not destroyed in the process. This could be a huge breakthrough for embryonic stem cell research. Amniotic stem cell research is not controversial, since the cells can be harvested without harming the embryo. Adult stem cell research is also not as controversial, so scientists are trying to find ways to "reprogram" adult stem cells to be more versatile and effective. Adult stem cell medicine is already being practiced to help regenerate blood for patients with leukemia (through bone marrow transplants), and some adult stem cells have been used to help ligament and tendon injuries in horses. Adult and amniotic stem cell research gets much more funding than embryonic since they are not as controversial.

Moving up from the cell, the next level would be the **tissue**, since tissues are made from groups of cells working toward the same function. **Organs** are composed of tissues and the whole **organism** is composed of different organ systems (cell → tissue → organ → organ system → organism). This chapter is an introduction to the different types of tissues, membranes, and cavities that hold the organs in the body. Once we have a general picture of where everything is located and what types of tissues and membranes are involved in the organ systems, we can start looking more closely at each system and try to understand their different components and functions.

Body Cavities: Function to protect and organize body's vital soft parts.
1. Cranial cavity with brain and cranial bones
2. Spinal cavity (vertebral) contains the vertebral column, spinal cord, and beginnings of spinal nerves
3. Chest cavity (thoracic) has three areas:
   - Pleural cavities, one for each lung, with serous membrane lining cavity
   - Pericardial cavity is around the heart
   - Mediastinum is the middle portion between the lungs and heart, including thymus, esophagus, trachea, and large blood vessels

The diaphragm (a thick sheet of skeletal muscle) separates the chest and abdominal cavity.

4. Abdominal cavity contains stomach, liver, pancreas, kidney, small intestines, and most of large intestines
5. Pelvic cavity is part of large intestines (colon), bladder, rectum, reproductive organs

Okay, now that we know generally where different parts are located in the body, let's look at what the organs and linings are made from. Every part of the body is composed of cells from one of the four types of tissues listed below. There can be many different TYPES of cells in some of the tissue types (connective, for example, has numerous types of cells), or they can be very specific like the nervous tissue, which only has two types of cells, but they have been grouped into these categories based mostly on how they perform in the body.

Four different types of tissues:
1. **Epithelial**—covers the outside of the body (skin), lines the inside of cavities and tubules, and makes up our glands.
2. **Connective tissue**—ligaments (connects bone to bone), tendons (connects muscles to bone, and muscle to muscle), cartilage (bone to skin and joints), fat (storage and cushioning), bone (structure), and blood/lymph (transports and protects).
3. **Muscle tissue**—three kinds: skeletal, cardiac, and smooth; provides movement and heat. Of the three types, skeletal tissue is the most abundant type and makes up 20–30% of our body mass.
4. **Nervous tissue**—all throughout the body, composed of neuron cells (to sense, process, and respond to stimuli) and glial cells (to support, insulate, repair, help nourish, and remove debris from neuron cells).

Muscle and nervous tissues will be discussed in their own chapters, but I will cover the first two tissue types here.

Epithelial tissue:

Functions include protection, sensation, absorption, secretion, excretion, diffusion, friction-reduction, and cleansing. (Sounds like a tongue twister!) Epithelium is sheet-like tissue with one side facing outward (either outside body, or inside organ or tubule). It can be **simple** with only one layer of cells (good for absorption, diffusion, and filtration), or have many layers of cells, **stratified** (good for protection and secretion from glands). There are eight basic types of epithelia based on their shape (flattened, cube-like, or columnar) and number of cells (simple or stratified). The shape and number determine the many different functions of epithelial tissues. Some epithelial tissues have cilia; for example, in your throat, nasal passages, and reproductive tract. The cilia function to help move things along, whether it be dust particles trapped in the mucus in your trachea, or moving the egg down the fallopian tube.

Glands develop from stratified epithelial tissue. How the secretions are released determines whether it is an endocrine or exocrine gland.

- **Endocrine glands** make hormones that directly enter extracellular fluid (there are no ducts); these hormones are then picked up by the circulatory system to be delivered to target cells in other locations. Ex: adrenal, pituitary, and thyroid glands.
- **Exocrine glands** release substances through ducts and tubules. Ex: salivary, mammary, mucus, and sweat glands.

Epithelial tissue makes up three types of **membranes** in the body, all of which have special functions:

Mucus Membranes: Digestive, respiratory, urinary, and reproductive systems are all lined with mucus membranes. Most of these systems (except the urinary) secrete mucus, which is helpful with lubrication, absorption, diffusion, protection, and getting the sperm to the egg! The digestive system also secretes saliva, which helps to start the digestive process. Most mucus membranes also function to absorb substances across the tissues.

Serous Membranes: Line body cavities and outer surface of organs, help anchor organs in place, and produce fluid to prevent friction between the organs.

Cutaneous Membranes: The protective durable outer layer of skin with three types of glands (oil, sweat, and wax—covered in the next section).

Connective Tissue: Ligaments, tendons, cartilage, bone, fat, lymph, blood, and specialized connective tissue membranes:
- Connective tissue is the most diverse tissue in the body.
- It connects, anchors, binds, and supports. Other critical functions include the storage of reserve energy; cushioning; blood cell formation; immune functions; tissue repair; transport of nutrients, metabolites, and hormones; heat generation; movement; and protection.

- Here's a reminder of some of the different types of connections: ligaments (connect bone to bone), tendons (connect muscles to bone, or muscle to muscle), fascia (connect muscles to muscles as well, and can also be around groups of muscles, nerves, or blood vessels), cartilage (connects bone to skin and joints and is also in the bronchiole tubes, intervertebral disks, ears, nose, and ribs). These types of connective tissue can secrete a non-cellular matrix composed of collagen fibers for strength and elastin for flexibility.
- Cartilage is the only connective tissue without a blood supply, so nutrients must diffuse through the cartilaginous matrix to reach the cells. This makes it harder to repair and form this tissue type. Hyaline cartilage is the most prevalent (at the ends of bones to prevent friction, and nose tissue). During development, the entire skeleton is composed of hyaline cartilage and then ossifies into bone. The other type of cartilage is called elastic (in epiglottis and ears). Cartilage is composed of specialized cells that produce a matrix of collagen and elastin fibers.
- Blood and lymph are fluid connective tissue (the matrix is not solid).
- Bones secrete a very solid ossified matrix made up of protein, water, calcium, and phosphorous salts. The bone cells can still communicate through channels called canaliculi, where nerves, blood, and lymph can travel.
- Adipose tissue doesn't have much of an extracellular matrix, but has cells specialized for storing lipids and can stretch as more lipids are deposited.

Connective Tissue Membranes: (makes up the other category of membranes)
<u>Meninges</u>: Special lining in cranial and spinal cavities. There are three meninges membranes (1. dura mater, 2. arachnoid mater, and 3. pia mater) with cerebrospinal fluid (CSF) between #2 and #3. The meninges and CSF are for shock absorbance and cushioning to the brain, and the CSF also helps with nourishment, support, and the removal of metabolic wastes from the brain.
<u>Sinovial membranes</u>: Line cavities of joints* and secrete lubricating fluid to prevent friction between bones and tendons.
*Joints occur where two bones come together.

There are <u>three types of joints</u>:
Synovial joints provide the most movement out of the three types of joints. They occur where synovial membranes secrete synovial fluid into the joint cavity. The fluid functions to lubricate and protect the bones. Examples: knees, ankles, hips, shoulders. ("Cracking joints"—synovial fluid has dissolved gases in it, and when it builds up and then is released, it makes a popping sound. This can help the joint feel more flexible and relaxed, but can also lead to decreased strength in the area, so don't crack your knuckles as a habit.)
Cartilaginous joints are where cartilage fills the cavity between the bones. Examples: in vertebrae and in ribs (less movement is possible in these joints).
Fibrous joints occur where there is no cartilaginous tissue connecting the bones. These types of joints don't allow much movement at all. Examples are in the skull and pelvis.

Tendons keep joints stable by holding the adjoining bones in proper alignment.
The other tissue types, muscles and nerves, will be covered in chapters 6 and 13.
Let's take a closer look at the **skin** since it is the largest organ of the body (weighing 9 lbs and covering 27 square feet on average in an adult), and is composed of both <u>epithelial</u> and <u>connective</u> tissues. The

outside layer is composed of the stratified epithelial tissue and is followed by the connective adipose tissues. Functions of the skin include defense against abrasion, chemicals, pathogens, and solar radiation; sensory functions; thermoregulation; and controlling water content and desiccation (keeps water in and inhibits the drying effects of air). Skin also produces vitamin D, necessary for proper bone development, and has excretion, secretion, and metabolic functions.

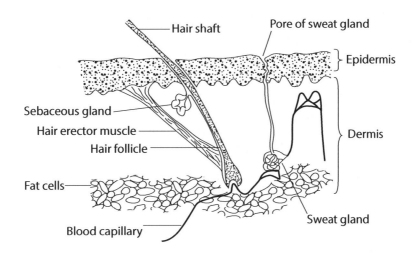

Layers of the skin:

The **epidermis** is the top layer, composed of stratified squamous epithelial tissue with a lot of keratin (a tough fibrous protein). The keratin is both inside the skin cells and between the tightly packed cells in order to provide skin its protective barrier quality. (The keratin helps to repel water and pathogens.) These cells are called <u>keratinocytes</u> and they make up 90% of the epidermis. The keratin replaces the cytoplasm as the cells move from the deepest layer of the epidermis, where they are being produced by cell division, to the top, where they die after 2 to 4 weeks of life. There are mostly <u>dead cells</u> on the outside protecting the living cells underneath. This layer does not have a blood supply, but some nutrients can diffuse from the dermal capillaries into the epidermis tissue to support the cells on their journey toward the surface; but the nutrients don't last long and this is the reason they die as they reach the top. Dead cells are sloughed off at a rate of roughly 35,000 cells a minute! (This accounts for most of the "dust" in your house, believe it or not!) As the old cells are sloughed off, the new cells rise to the top. We replace our entire top layer of skin every 4 to 6 weeks.

There are also <u>melanocytes</u> (8% of epidermis) in the bottom of the epidermal layer. These cells produce the pigment melanin. How much melanin a person has determines the color of their skin, and depends on a number of different genes. The melanin pigment is transferred to different cells called keratinocytes, which make up the majority of the cells in the epidermal layer. While the melanin gives the skin its color, the keratin in these cells provides both the tough and waterproof qualities of the skin. (There are no melanocytes in the palms of the hand or soles of the feet, so these are always a light color even in darker-skinned people.) Melanin also protects the skin from UV radiation by absorbing the UV radiation before it can penetrate deeper into the skin, thus damaging the DNA in cells. Too much sun exposure can lead to skin cancer, however. Pockets of melanocytes cause freckles and moles (also hereditary). If you have a first-degree burn, it is only within the epidermal layer. A second-degree burn would be all the way to the upper layer of the dermis, and a third-degree burn is throughout both epidermis and dermis layers and can be life-threatening.

Fingerprints in the outer epidermal layer of the skin are genetically determined ridges on fingers. Identical twins do not have identical fingerprints, however, despite having the same DNA. This is because the swirls and patterns develop differently as the embryos are formed in utero, so no two fingerprints are the same. The epidermis also has defensive cells called <u>Langerhans cells</u>. These are phagocytes that can engulf bacteria and viruses within the skin. The fourth, and least numerous, of the epidermal cells are called <u>merkel cells</u>. These cells, located in the lower layers of the epidermis, are in contact with nerve cells coming up from the dermal layer and help with perceiving sensations. The epidermis is connected to the next layer, which is the dermis, by the **Basement Membrane**, which is filled with proteins and polysaccharides helping to bind the two layers together.

The **Dermis** layer is mostly connective tissue, and is more flexible due to the collagen (protein) and elastin fibers in this layer. These proteins allow the skin to stretch and retract back to its original shape (helps with facial expressions and is important as we gain and lose weight or during and after pregnancy). This layer also has glands, hair follicles, blood, and nerves. The dermal layer does not shed like the epidermis (this is why tattoo ink is injected into this layer). As we age, the collagen and elastin in this layer degrades, which leads to wrinkles as the skin sags in areas that used to be filled with this tissue. This happens the most in areas that have been frequently creased due to smiling, frowning, squinting, whistling, etc. The dermal layer supports the epidermis and there are plenty of capillaries to nourish the cells in this layer. Millions of sensory neurons are conducting signals from the skin, and glands produce oil and sweat to help with lubrication, protection, and thermoregulation.

Three types of glands located in the skin (from epithelial tissue)

1. **Oil (sebaceous) glands** are found everywhere except on the palm of the hand and soles of the feet. <u>Sebum</u> (oily substance from fats, proteins, and salts) is secreted and functions to lubricate the skin and hair and protect from bacterial growth. During puberty the androgen hormone levels increase and trigger the release of more sebum, which can lead to acne. **Acne** is basically excess sebum and dead skin cells that clog sebaceous glands that open onto hair follicles. If the follicle is obstructed by sebum and dead cells, it results in a whitehead, versus if the follicle is clogged by sebum and melanin, it is a blackhead. A pimple will form when the follicle ruptures and bacteria infect the area. High-sugar diets can make acne worse in some people by triggering more inflammation in the epidermis. Acne is not due to having poor hygiene, since follicles are being plugged from below. However, washing your face with hot water can help unplug clogged follicles. There are many medications that can help as well, such as Benzoyl peroxide to kill bacteria living in follicles, Retin-A™ to thin the epidermis so that dead cells don't clog the follicles, and other retinoids that poison and shrink the sebaceous glands to reduce the amount of sebum produced. (These should only be used for very severe acne. They can cause birth defects, so patients using retinoid medications need to be sure to use safe birth control concurrently.)

2. **Sweat glands** are very important for thermoregulation, and we have a lot of them—2.5 million! Sweat is 99% water, with the other 1% being salts, sugar, ammonia, lactic acid, vitamin C, and the metabolic waste urea. We eliminate 30% of our waste through the skin! (Other waste is processed by our lungs [CO_2], kidneys, GI tract, and liver.) When the body temperature rises above 98.6 ° F, the hypothalamus in the brain triggers sweat glands to start secreting sweat (perspiration) through the skin's pores. The

liquid turns into a vapor when it comes into contact with the air and this evaporation of water cools the body down by absorbing energy in the process. Why does sweat smell if it's 99% water? It's actually bacteria on the skin mixed with the sweat that causes the smell. (Odors from breath, feet, and feces are also due to bacteria—the smell often depends on how many and what kind of bacteria you have. It can also depend on what foods you usually eat. Some bacteria produce sulfur compounds as they multiply—super stinky!) Puberty hormones trigger the production of more sweat, so that is when people usually start wearing deodorant. It is also the time when hair starts to grow under the armpits, which helps to trap more of the sweat and bacteria and therefore cause more of the smell. When you sweat a lot, it is good to drink plenty of fluids to rehydrate your body. Since you also lose salts in sweat, it is important to replace your electrolytes after a big workout as well (sports drinks do this).

3. **Wax glands** are modified sweat glands and are located only in the external ear canal. The wax produced here protects the ear by trapping particles or pathogens entering this orifice.

The **Hypodermis** is technically not part of the skin. It is a subcutaneous <u>connective tissue</u> layer that anchors the skin to muscle and bones and is composed of mostly adipose connective tissue, jelly-like fat, which functions for cushioning and insulation. (The hypodermis contains 50% of our body fat.) We lose water and fat in this layer as we age, which also adds to the wrinkling process of the skin. This layer also has many macrophages and blood vessels, which feed into the dermal layer. In babies, this layer covers the entire body for protection, cushioning, and thermoregulation. The fat gets redistributed during the aging process and certain areas accumulate more than others. Generally men have more fat in the tummy/waist area, and women store their fat more in the thighs, breasts, and hips. Women generally have a thicker hypodermal layer, giving women a curvier appearance than most men. Where fat is stored the most can also be genetically determined. (<u>Liposuction</u> is a procedure used to suck out fat cells in this hypodermal layer. It can decrease a person's weight by only a few pounds, but can smooth out bulges and remove the fat-storing cells so that the amount of lipid that is stored in the area in the future is decreased. It can be dangerous in some cases, however, causing blood clots to travel to lungs.) A better way to reduce fat is to exercise, maintain a healthy diet, and not overeat.

<u>Hair and nails</u> also provide protection and insulating functions to the skin.

- Nails are modified skin tissue, hardened by keratin (a very tough protein), and can grow continuously. They are embedded in the skin, and the part that you see is all dead cells. Other species of animals have this keratin tissue in other forms as well, such as hooves, horns, whiskers, claws, fur, shells, scales, feathers, and quills!
- Hair is modified dead skin cells filled with keratin, packed in a column to provide channels for oil and sweat glands. Also very important for protection (the hair on your head protects you from UV radiation, the hair in your nostrils protects you from particles entering your respiratory system, etc.)
- Muscle cells are attached to hair follicles (goose bumps occur in response to fear or cold when muscles contract). We have it as a leftover adaptation from when we had more hair on our skin. Hair standing straight up helps with insulation and is also used to threaten others in tense situations. (Think of a cat with its hair on end.)
- Hair, fingernails, and toenails all develop from the skin epidermis and all have lots of keratin.
- An average human head has about 100,000 hairs, and we lose about 100 per day. (It seems the more you have, the more you lose.)

- Each hair grows out of its own follicle and is active about 2–6 years before it falls out. Luckily they don't drop out all at once, but are staggered. You continually replace the hair that is lost, although the rate drops off as you age (most elderly people have thinner hair).
- Hormones, nutrition, aging, and other factors affect hair growth as well as rate of hair loss.
- The shape of the hair follicle determines whether it is straight, frizzy, curly, etc., and is hereditary (as well as the hair color).

Threats to Skin

- Smoking cigarettes is very damaging to the skin.
- Toxins in tobacco break down collagen, elastin, and other proteins in the skin and can cause deep wrinkling.
- Constant pursing of the lips to smoke enhances the wrinkles in the mouth area.
- Nicotine's constriction of blood vessels stops the flow of nutrients to cells, making skin cells die faster than usual.
- Secondhand smoke can also affect skin, often causing eczema or atopic dermatitis, especially in children of smokers.
- UV radiation can break apart DNA nucleotide sequences or double helix bonds. DNA can be repaired but sometimes the sequence is not the same and the mutation can lead to skin cancer.
- The Sun has UVA and UVB wavelengths. The UVB stimulates vitamin D synthesis by skin cells. It also causes the skin to burn more quickly. The UVA is actually more dangerous to our body. It weakens the immune system, which makes it harder to fight off cancer. It also can destroy folic acid, which is needed for DNA synthesis.
- Glass windows blocks UVB radiation so we need to expose the skin directly to the sun at least a few times a week to get enough vitamin D. Glass does not block UVA radiation so you can still get sunburned through a window and have the aging and cancerous effects from too much light through glass. It can also cause macular degeneration.

UV radiation can also activate proto-oncogenes, which cause **skin cancer** (three types):

1. Squamous cell carcinoma occurs in newly formed skin cells in the epidermal layer as they flatten and move toward the surface.
2. Basal cell carcinomas occur in the deeper layer of the epidermis where the skin cells are dividing (this is the most common type of skin cancer).
3. Malignant melanoma (the most dangerous of the skin cancers, but also the least common) develops in your melanocytes and can spread through your lymph to organs all over the body; it causes death when it affects the functioning of another organ such as the lung. This type occurs in lighter-skinned people much more frequently than in people of color. It is also the most common type among kids. People are 50% more likely to develop skin cancer later in life if they have had 5 to 6 strong sunburns as children. Kids should especially never be taken to tanning salons, a major cause of melanoma. (California Governor Jerry Brown recently signed a law to prohibit minors from using tanning salons—hurray!)

- Skin cancer is not something to take lightly (no pun intended!) Two million people develop skin cancer in the United States annually and 12,000 die due to this cancer every year.

- If you notice a growing or raised freckle, go have it checked out, since it might be an indicator of skin cancer. (Cancer Society **"ABCD skin cancer check"** includes looking for **a**symmetrical shape of mole or freckle, unclear **b**orders, **c**olors on melanomas, and **d**iameters greater than .2 inches [5 mm].)
- Remember that the skin needs a certain amount of UV radiation to trigger our bodies to produce vitamin D (from cholesterol). Vitamin D helps us absorb calcium, which is important for bones, muscles, nerves, etc. Vitamin D may also protect against colon cancer, breast cancer, and multiple sclerosis. Dangers of sun exposure can sometimes outweigh the benefits of getting the vitamin in this way (you can also get vitamin D in foods and supplements), so be sure to have fun in the sun, just limit your exposure and …

Wear sunscreen! Most sunscreens block the UVB (which causes skin to burn), but they don't always block the UVA. Try to use a sunscreen with both types that provides 30 SPF protection. Avoid sunscreens with Retinyl palmitate (vitamin A), Oxybenzone, and sunscreens that are sprays or powders. These and other dangerous ingredients may do more harm than good. You still need to be careful with your sun exposure even when you are "gooped" up. And don't be fooled by the "waterproof" label: still reapply the sunscreen after being in the water. Also reapply every two hours, avoid the sun between 10 am and 2 pm, and protect the skin with clothing, a hat, and glasses when possible. The Environmental Working Group (EWG) recommends sunscreen with zinc oxide (blocks both wavelengths), titanium dioxide, avobenzone, or Mexoryl S. Wearing sunscreen not only protects you from skin cancer but it is also the best way to slow the aging process in the skin.

Sunburn: UV radiation can destroy the outermost layer of the epidermis (surface of skin). This causes blood vessels in the dermal layer to dilate in response to the damaged tissue (to aid in repairing the area by bringing more O_2 and nutrients), and therefore the skin appears red. A new layer of cells is generated and pushed up from below, and the burned skin peels or sloughs off.

Tanning is due to the fact that the Sun triggers melanocytes to increase melanin production, but in the long run this is more harmful than beneficial. Although melanin can protect against UV radiation, tanning is often exposing your cells to more than the melanin can protect against and can lead to cancer. Long-term tanning also breaks down elastin fibers and makes it hard for the skin to replenish itself. This causes the skin to lose its resilient quality and leads to more wrinkles.

- Definitely don't go to tanning salons! Tanning salon bulbs can deliver five times the Sun's UV radiation! They use the UVA wavelength because this is the kind that doesn't burn the outer skin, but it is actually the more deadly radiation since it causes melanomas. One million people use tanning salons every day. Rates of melanoma skin cancer have doubled since 1975 in white women ages 20–49 due to tanning salons.
- Light-skinned people are generally more at risk for skin cancer since they have less melanin than darker people. Albinos lack an enzyme that is necessary for the production of melanin and therefore definitely need to wear sunscreen because they don't have any melanin protection at all. (There was an albino kangaroo at the San Francisco Zoo and they would apply sunscreen to this animal every day!)
- Darker-skinned people have less risk of skin cancer, but melanin lowers vitamin D production, so they need to be sure to get enough in their diets.

Chapter 6

Muscular System

Half of the human body is composed of muscle tissue! There are three types of muscle tissue: **skeletal**, **smooth**, and **cardiac**. Each of these types has different muscle cells. The functions of muscles are to provide movement and for thermoregulation (to produce heat). The movement can be voluntary, such as our skeletal muscles moving our bones, or it can be involuntary, where we don't even have to think about the internal movement of substances, such as the blood coursing through our bodies by the contraction of our heart muscles or food moving through our digestive system by the contraction of our smooth muscles lining our digestive tubules. The contraction of muscles is under the control of the nervous system, so the two systems are closely connected, as you will see.

Skeletal muscle is the most abundant tissue type in the body and makes up 25–30% of the body mass. (This is an average percentage for healthy adults; some people have much more muscle, especially body builders!). Skeletal muscles make up the voluntary muscular system as opposed to the cardiac and smooth muscles, which are both involuntary. There are approximately 640 skeletal muscles in adults, and they are attached to bones by tendons. When they contract they force the bones to move. Ligaments attach bone to bone and cartilage attaches bones to joints (between bones) and skin, so when the muscles move, all of these connective tissues are forced to move as well. Some work in groups or pairs, and many are antagonistic; for example, when the bicep contracts, the tricep relaxes.

Muscle cells are bundled together in connective tissue, and this package is what we call a muscle; each different muscle is considered an organ. Skeletal muscles are highly organized, with cells lying parallel to one another, and they are multinucleated (have many nuclei in one cell). They also have many mitochondria to produce a lot of ATP; energy is needed for both contraction (when they get shorter) as well as relaxation (when they go back to their resting length or even elongate past that stage). The contraction of muscles is what causes movement. Skeletal muscles appear striated due to the banding of actin and myosin proteins, which help in the contraction process. Some skeletal muscles we don't have to think about to make them move, like the diaphragm muscle, which will move with every breath we take. Luckily we don't have to consciously remember to breathe. That is a reflex controlled by the nervous system. There are other reflexes that will force muscles to move without us consciously thinking about the movement (e.g., the knee jerks if hit in the right spot), but they are still part of the "voluntary" skeletal muscle system.

Now let's look at how skeletal muscles work:

- Skeletal muscles are stimulated by motor neurons.
- When the nerve impulse travels down the axon to the presynaptic knob of the motor neuron, it triggers the release of the neurotransmitter **acetylcholine** into the space between the motor neuron and the muscle cell.
- The space between a neuron and muscle cell is called the **sarcolemma** (also known as the synapse or neuromuscular junction).
- Acetylcholine (ACh) is a neurotransmitter, which binds to receptors on the muscle cell motor end plate and causes a depolarization of the membrane. This stimulates the T-tubule system in the muscle cell to signal the **sarcoplasmic reticulum** to release calcium ions into the cytoplasm of the **sarcomere**, the muscle unit composed of the globular proteins actin and myosin. (Many sarcomeres are bundled together in myofibrils.)
- ACh is then removed from the sarcolemma by the enzyme **acetylcholinesterase** (AChE), which can degrade 25,000 molecules of ACh per second! The particles get recycled to make more ACh for subsequent nerve impulses to transmit, but if they weren't removed, the muscle cells would spasm with constant contraction.
- Both actin and myosin have binding sites where they can attach with one another, but it happens only during muscle contraction.
- The thin proteins are called actin, and these are attached to the Z-line toward the outside of the sarcomere, whereas the thick proteins, myosin, are lined up in the center of the sarcomere between actin proteins.

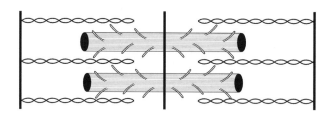

- When a muscle is relaxed, the actin has inhibitor proteins bound to the binding sites, which inhibits myosin from attaching to these sites.
- When calcium enters the sarcomere, the Ca^+ causes the inhibitors attached to the actin to leave the binding sites. (In this text, when I write "Ca^+", I'm referring to calcium ions, which technically should be written as Ca^{++} or Ca^{+2} since they are missing 2 elections. However, for simplicity's sake, I will always just refer to them as Ca^+.)
- Myosin has globular heads that can fit into the active sites of actin. Once the inhibitor proteins are removed from the active sites, the myosin can create cross-bridges that attach to the actin active sites, and pull the actin to slide from one globular head to the next toward the center of the sarcomere. This

contracts the whole sarcomere as the actin is sliding the Z-lines toward one another. This is known as the "**sliding filament theory**."

- ATP provides energy for the myosin to go from one active site to the next, also referred to as the actin "walking" along the myosin.
- Each sarcomere contracts independently, but they are all united by their Z-lines. (Millions of sarcomeres are lined up in one muscle cell, and many muscle cells are bundled within connective tissue to make a muscle. One motor neuron can innervate many muscle cells at once to contract the muscle.)
- When all the sarcomeres are contracting, the entire muscle is contracted.
- As the sarcomeres contract, the muscle thickens.
- One side of the muscle (the "origin end") stays stationary, while the other side (the "insertion end") moves during the contraction.
- When Ca^+ levels are high, the muscle will contract, and when the Ca^+ levels are low, the inhibitor proteins reattach to the actin binding sites and the muscle will relax. ATP is also used to pump Ca^+ out of the cell during the relaxation phase. Ca^+ is also brought back to the sarcoplasmic reticulum during the relaxation phase.
- Energy is also needed to break the actin-myosin bonds during relaxation of sarcomeres, so muscle cells need a constant supply of glucose and oxygen in order to make ATP (by cellular respiration in the mitochondria).
- Only 15% of the energy created during muscle contraction is needed for the mechanical process; the rest is given off as heat, which is important for maintaining a constant internal body temperature (thermoregulation).

When you die, your body stops producing ATP, so there is no energy to break the bonds between myosin and actin that is contracted and no bonds to pump Ca^+ out of the cells. The membranes become more permeable to Ca^+, so actin and myosin form more cross-bridges that can't be broken after death. Rigor mortis, "stiffness of death," is due to this contraction of muscles. (It doesn't occur immediately upon death, but sets in anywhere from ten minutes to twelve hours after death, depending on the temperature and other circumstances. The body doesn't loosen again until the dead tissue starts to decompose 24–60 hours later.)

When you exercise you need plenty of oxygen and glucose for cellular respiration. If you run out of glucose, you start breaking down the glycogen the liver has stored. If you run out of glycogen, then you can use fats. If you run out of fats, the body will start breaking down muscle, which is not good since you can't regenerate these cells once they are gone. If you run out of oxygen during strenuous exercise, your body can switch to **glycolysis** (anaerobic respiration), which produces ATP (not as much as cellular respiration), but also produces lactic acid. Getting a cramp during strenuous exercise is often a sign that you are producing lactic acid. On the other hand, feeling "sore" after exercising is usually caused by small tears in the muscles that you were using. As your body repairs the area, it enlarges these cells, making these muscles stronger. It will be easier to do the same exercise the next time around.

When we reach middle age, we start to lose roughly 6.5 pounds of muscle every decade. This means we don't burn as many calories when we exercise and we start to store more fat. Our fat metabolism slows down over the course of our life because we lose a lot of mitochondria as our muscle mass goes down—mitochondria is where we convert glucose to ATP. If we're not converting enough glucose to energy, we convert it to fat instead. Staying active by participating in aerobic exercises like jogging, swimming, walking, dancing, sports, etc., is very important for MANY reasons.

Benefits of exercise:

1. Keeps muscles active and healthy and increases their efficiency.
2. Burns calories, and helps to maintain a healthy weight. You actually need less energy for metabolic functions if your body is toned versus if the body has excess fat.
3. Improves muscle coordination.
4. Develops new blood vessels and increases the amount of **myoglobin.** (Myoglobin is the pigment used to deliver oxygen in muscle tissue rather than hemoglobin.)
5. Develops more mitochondria in muscle cells (remember, this is where ATP is made).
6. Increasing the number of mitochondria in cells due to exercise will also increase your energy levels and the release of heat, important for thermoregulation.
7. Increases bone density and prevents bone loss.
8. Makes heart muscles stronger.
9. Pumps blood more efficiently, delivering oxygen and nutrients more efficiently.
10. Reduces blood pressure.
11. Reduces bad cholesterol and increases good cholesterol in blood (muscle contraction stimulates the release of lipoprotein lipase, which metabolizes triglycerides and helps to make HDL—the good cholesterol).
12. Better digestion and elimination of wastes.
13. Having healthy, strong muscles will give you a faster metabolism.
14. Reduces stress—very important, since stress weakens your immune system.
15. Helps with sleep—critical for health, growth, mental clarity, memory, etc.
16. Improves mood and increases alertness by stimulating the release of endorphins, serotonin, and/or dopamine.
17. Very important for your brain! People who excercise have less brain shrinkage during their life and exercise enhances cognitive flexibility and function. (Studies show that kids who get to exercise regularly during the day perform better in school.) It also makes people more receptive to learning.
18. Increases your chances of a longer healthier life!

There are three main types or exercise: **cardiovascular**, **strength training**, and **flexibility**. All three are important to have in your weekly routines. Doing some sort of cardivascular exercise for <u>30 minutes, 5 times a week</u> is recommended not only to maintain a healthy weight and cardiovascular system, but also for boosting metabolism, energy, mood, and alertness. Strength training (recommended at least <u>20 minutes two times a week)</u>, is especially helpful with brain functions and increases clarity and ability to plan and multitask. Flexibility exercises like yoga, pilates, tai chi, and stretching help to lengthen tendons and muscle fibers, reduce anxiety, and promote better sleep, flexibility, and balance. Muscles that are more flexible will increase in size and become stronger, which boosts circulation and metabolism.

Strength training increases muscle size (hypertrophy) more than aerobic exercise, but aerobic exercise has more health benefits. If weight loss is the main goal the best exercise is called "interval training." This is when you do bursts of intense aerobic activity between moderate aerobic exercise. The bursts will help burn more fat but keeping most of the exercise moderate will help you exercise longer in general. You also burn more fat if you eat before you exercise, according to new research. Muscle cells start to <u>grow</u> once you exert more than 75% of their maximum force. (You don't make MORE muscle cells. In fact, even if you lose muscle cells, the body will not regenerate these cells.) The amount of muscle cells and kinds of skeletal muscle cells you have are genetically determined. The two types are red (slow-twitch) fibers, which are better for aerobic exercise since they don't fatigue as quickly (they have lots of blood vessels, mitochondria, and have a special protein myoglobin to carry oxygen, needed to make a lot of ATP), and white (fast-twitch) muscle fibers, which enlarge more with weight training, have higher glycogen reserves, and provide short bursts of extreme energy and strength. These fibers have fewer capillaries and less myoglobin and mitochondria, but can contract faster than red fibers; they just can't sustain the contractions for very long (good for sprinting).

Anabolic steroids are drugs that mimic testosterone, which stimulates muscles to increase in size by stimulating protein formation and reducing the rest stage between contractions. Many athletes take these drugs to enhance performance and muscle size without realizing the numerous possible side effects (at least 70!). Anabolic steroids can inhibit the natural production of testosterone, and can cause acne, cancer, strokes, stunted growth, liver damage, high blood pressure, heart attacks, smaller testes, low sperm count, etc. (There are other medications made from steroids as well, to relieve pain, rash, inflammation, and asthma, but these are not testosterone mimics and therefore do not have the same side effects.)

Muscular Problems

Muscle fatigue occurs when you run out of glucose and glycogen for cellular respiration. Your muscles start to "fatigue," which means they don't contract as much. (This is like when a marathon runner "hits the wall"—when energy-storage molecules are running low, they suddenly lose energy. Most marathon runners will "carbo load" before a race by consuming a lot of pastas, etc., to load up on glucose/glycogen for the race.)

Muscles cramp when you don't have enough oxygen for cellular respiration. Your cells switch to glycolysis (which is anaerobic respiration without oxygen), to produce ATP. The byproduct, however, is lactic acid, which can make your muscles feel cramped. (It's good to let your muscles rest when they are really cramped so that your body can degrade the lactic acid that builds up from anaerobic respiration—this generally takes about an hour after intense exercise.) Muscles can also feel cramped from low calcium, magnesium, potassium, or sodium. Eating a banana (high in potassium) before working out will sometimes help prevent cramps and also provide plenty of sugar for cellular respiration so that muscles don't fatigue. Stretching can also help relieve cramps; be sure to drink plenty of water before, during, and after strenuous exercise.

Muscle strain (or "a pulled muscle") is when the muscle fibers, or the connective tissue associated with the muscle, tear due to overstretching. A muscle strain is different from a **sprain**, which is when you injure a joint by over-stretching the ligament associated with the joint. (Sprains are relatively common injuries, especially in the wrists and ankles.)

Sore muscles are often caused by a strenuous workout that stretches or damages muscles that are unaccustomed to the activity. You usually don't feel sore until 24–72 hours after the activity. Your body will repair the small tears in the muscle tissue as well as build up even more tissue than before. This makes the muscles bigger and stronger and better able to handle the same strenuous activity the next time you do it. When muscles are very sore, applying heat will help to relax the muscles and bring more blood to the area to help repair tissue damage, although applying cold packs are also helpful to relieve pain (since it numbs the area) and inflammation. Alternating between heat and cold is often recommended for muscle strains or sore muscles.

Muscular dystrophies are genetic disorders where muscle cells break down, causing muscles to shrivel and weaken. Severe cases often lead to the use of a wheelchair.

Bacterial infections can affect muscles cells too. Botulism is due to a bacterium that produces a neurotoxin that stops motor neurons from releasing acetylcholine. If you don't release acetylcholine you also don't release Ca^+. Muscle contraction cannot occur without the calcium from the sarcoplasmic reticulum, and the muscle becomes paralyzed. Once the heart or respiratory muscles become paralyzed, death occurs. (It can be stopped if treated in time with an anti-toxin.) Small doses of this neurotoxin are used in botox injections to paralyze muscles not only for cosmetic purposes but also to stop muscle spasms in people affected by disorders such as cerebral palsy. Tetanus is also due to a neurotoxin from a bacterium that causes muscle paralysis, the first sign of which is usually "lock jaw." The bacterium enters a puncture wound contaminated by animal waste or saliva, rust, or dirt. (Tetanus incidents are usually due to animal bites, rusty nails, or large splinters.) Once the neurotoxin reaches the spinal cord, it blocks the nervous system from sending any messages and therefore muscles don't contract. Once the heart and respiratory muscles stop contracting, the person dies. Most children now get tetanus vaccinations to avoid this disease. It is important to get a booster shot every 10 years as an adult as well. Tetanus still kills approximately 200,000 people per year worldwide.

Tics, spasms, and charlie horses are abnormal contractions of muscles that can be caused by vitamin deficiencies, electrolyte imbalances, hormone changes, and stress.

Muscle spasms can also be caused by pesticide exposure. Pesticides are chemicals used to kill insects on crops and gardens. However, they affect humans as well as insects, especially when exposure is high (for instance in farm laborers). Heavy exposure to pesticides kills roughly 500,000 people per year and causes numerous other health problems, especially associated with the nervous system. Some of the chemicals in pesticides disrupt the enzyme acetylcholinesterase, which breaks down acetylcholine. If acetylcholine doesn't break down in the sarcolemma of muscle cells, it accumulates and causes continual muscle contractions (spasms).

Sarcoma refers to different types of cancer affecting tissue derived from mesenchyme (like muscle tissue). Cancer in muscle is relatively rare.

There are two other types of muscles; both are **involuntary** with steady contractions all the time:

2. **Smooth** muscles
   - Found in linings of vessels, blood, lymph, urinary tract, digestive, and reproductive systems, they are not connected to bones.
   - Can be arranged in different layers (in the digestive system there are circular, longitudinal, and oblique smooth muscles in different locations). Alternate contractions cause a wavelike movement called **peristalsis**.
   - Sphincters are also made of smooth muscles and open or close either by autonomic or voluntary control. Contraction of the muscle closes the sphincter and relaxation opens the sphincter.

- Smooth muscles are triggered by neurons.
- Smooth muscles are not striated and only have one nucleus.

3. **Cardiac** muscles
   - Found only in the walls of the heart. They are the most durable muscles in the body.
   - Contractions pump blood through the heart and out into the vessels.
   - They are striated like skeletal muscles, but they only have one nucleus.
   - They are the only type of muscle that can contract without nerve impulses.

We will go over the heart in more detail in Chapter 8.

Chapter 7

Skeletal System

There are three types of skeletal systems in the animal kingdom:

Exoskeleton (on the outside of the body) is usually composed of hard mineralized substances such as chitin, sclerotin, and calcium carbonate. Examples include all of the arthropods: spiders, insects, crustaceans, centipedes, millipedes, etc.

Hydrostatic skeleton is made from fluid inside the body to help with support and movement. Different hydrostatic organisms use different fluids for their skeletons, such as water, mesenchyme, and hemocoel. Examples include starfish, anemones, earthworms, etc.

Endoskeleton is also on the inside of the body, but instead of fluid, the endoskeleton is composed of hard mineralized bone tissue making up a framework for the body where the muscles are attached. The main functions of the endoskeleton are protection and movement. Examples of organisms with endoskeletons: amphibians, fish, reptiles, birds (they have hollow bones, so they are light enough to fly), and mammals.

The mammalian skeletal system is composed mostly of two types of connective tissue, bone and cartilage. This is because the cartilage connects the bone to skin, and we have a lot of skin (27 square feet) and a lot of bones. The number of bones in your body depends on how old you are. As babies, we start out with 300 bones. They are soft bones at this stage, made of cartilage and easily pliable, which helps us to get out of the birth canal. They have all hardened completely into bone by the time we are 14 months old. As we develop, certain bones fuse together, and by the time we are 20 years old, most people have 206 bones. (However, some people have an extra rib and some have an extra bone in the arch of the foot so the total can be 207 or 208 depending on your genetics.) Our body is composed of 18% bone.

Here's just a reminder of some of the other tissues we talked about in Chapter 5 and how they are associated with bones:

Cartilage comes in so many different forms that it is hard to define. It supports other tissue types more than connecting them. We have some cartilage that stops our bones from grinding together and some that connects skin to bone; and there is cartilage in our respiratory tract, ears, nose, ribs, and knees. It is unusual in the fact that it doesn't have any blood vessels to nourish the cells so it heals and grows very slowly.

Ligaments connect bone to bone to form a **joint** (the site where two bones come together). There are different types of joints. Some are linked by ligaments, fibrous connective tissue, or cartilage and they are also different depending on whether they have the presence or absence of a sinovial cavity around the joint. Joints that have a sinovial cavity allow greater mobility than other types of joints; examples of sinovial joints are elbow, shoulder, knee, and hip joints. **Tendons** join bone to muscle as well as muscle to muscle. They also keep joints stable by holding adjoining bones in proper alignment.

Functions of Bone:

- Protection of organs—skull protects brain, ribs protect lungs and heart, etc.
- Support—provides the framework for the body; vertebrae give us upright posture.
- Blood production—marrow produces red and white blood cells and platelets.
- Movement—with the help of muscles, tendons, ligaments, joints, and cartilage, the bones help manipulate the body's movements.
- Mineral storage—especially calcium and phosphorus. They are both held in bone tissue and released when needed by the body for other functions.
- Acid–base balance—bone helps to buffer blood against pH changes by accepting or releasing alkaline salts.
- Detoxification—by storing heavy metals and other elements that can gradually be released for excretion.
- Shock absorbance—think of jumping off a wall; it's your bones that help to keep you from crumbling into a heap!
- Hearing—three little ear bones are critical for sound transduction.

Basic Bone Structure:

Compact bone accounts for 80% of the total bone mass in the adult skeleton. This smooth, solid, white, hard outer layer of bone has very few gaps in the tissue. Compact connective tissue is composed of three basic cell types: osteocytes, osteoblasts, and osteoclasts. **Osteocytes** are the most abundant cells in compact bone and are made from **osteoblasts**, which are bone-forming cells. These secrete a matrix all around themselves to provide the support and rigidity of bones; once they are trapped in this matrix, they are considered osteocytes and have cytoplasmic extensions to exchange nutrients and waste. The area the osteocytes occupy is called the lacuna. The osteoblasts are responsible for secreting the collagen and minerals of bones. The number of osteoblasts decreases during the aging process, which causes decreased bone density and strength. The **osteoclasts** are bone cells that break down bone tissue to release Ca^+ into fluids. Collagen is the most abundant protein in the human body and makes up 90% of the organic weight of bone. Elastin, other proteins, and some carbohydrates are also secreted by osteoblasts in the bone-building process. The minerals zinc, copper, sodium, calcium carbonate, and calcium phosphate are also in the matrix as well (bone tissue is hard, mainly due to the calcium salts). There are channels called **osteons** for vessels and nerves to deliver nutrients, gases, and messages to the bone tissue. Both osteoclasts and osteoblasts are under hormonal regulation to help with bone remodelling, which will be discussed below.

Spongy bone (also called cancellous bone), accounts for the other 20% of bone. It is located inside the compact bone and has large cavities filled with marrow. Spongy bone is lightweight, weak, and soft compared to compact bone. It is also less organized and lacks osteons. It has ten times the surface area of compact bone, allowing room for blood vessels and marrow. There are two types of marrow. **Red marrow** makes new red blood cells, white blood cells, platelets, skin cells, and other regenerative cells from stem cells. Red marrow is located in the hip bones, ribs, shoulder blades, breast bone, collarbone, and spinal vertebrae. Most other bones have **yellow marrow**, which stores fats in the form of triglycerides (the precursor for energy) and minerals; both can be released into the bloodstream when necessary. Yellow marrow is mostly in long bones. Babies have all red marrow.

As we age, more and more marrow converts to yellow marrow, and by the time we are adults, 50% of marrow is yellow. Yellow marrow can be converted back to red marrow if the body has a blood shortage (sometimes due to anemia or injury). The kidney will detect the blood shortage (by monitoring oxygen levels in the blood) and stimulate the conversion of yellow marrow back to red marrow if the oxygen levels are too low. (This marrow conversion often occurs in the femur, the long bone in the thigh.) The kidney will release the hormone **erythropoietin** (EPO), which stimulates stem cells in bone marrow to make more red and white blood cells.

Bone is constantly replacing itself throughout life; this is called **bone remodeling**. Remodeling of bone keeps the tissue resilient and healthy. (Bones may stop growing in length at maturity, but they are constantly changing in shape and density as calcium is added and taken away, depending on the circumstances in the body. There is a balance between cells that **b**uild (osteo**b**lasts) and that break down or **c**rumble (osteo**c**lasts) bone. Osteoblasts, as mentioned above, secrete collagen and elastin, some carbohydrates, and other proteins, and then pull calcium and minerals from the bloodstream into this matrix to create bone tissue. The calcium hardens the new bone and it is "ossified." Ossification refers to the process of osteoblasts' building bones. Once bone cells are ossified, they are called osteocytes. (Bones grow longer and thicker as osteoblasts add matrix to the exterior of bones and then become osteocytes at ossification.) Bones are either formed within mesenchyme tissue or within cartilage (more common). The whole process going from the original osteoblasts to the osteocyte-and-matrix tissue called ossified bone can take roughly three months.

Exercise and stress on bones actually helps to stimulate bone building. People who exercise have stronger, denser bones than couch potatoes have. (So for those of you who exercise a lot—which is great—but can't figure out why you still weigh as much as you do, take into consideration that muscle is denser than fat and that bone density is heavier with a healthy amount of exercise.)

On the flip side, osteoclasts (large cells that bind to bone tissue) will release acids and enzymes to help break down the bone-matrix tissue, releasing minerals like Ca^+ and phosphorus into the bloodstream. This is a very important job, because the body needs calcium for so many different functions. Bones contain our largest supply of calcium in the body. **Calcium is important** for muscle contractions, nerve impulses, blood clotting, and as co-enzymes for certain reactions. Calcium also helps nourish the developing fetus, and is required for lactation (milk production and delivery). Vitamin D helps the body absorb calcium from the digestive system into the blood. (Sun exposure triggers the skin to produce vitamin D.) Vitamin D also helps the kidneys absorb calcium so that it is not lost in the urine. Vitamin D deficiencies can lead to rickets and osteomalacia—both of which are bone-weakening diseases.

Always replenish Ca⁺ and vitamin D with foods and drinks (milk, orange juice, cheese, and other dairy products, plus certain fruits and veggies are all great sources of calcium and vitamin D). Caffeine lowers the amount of calcium deposited into bones, and carbonated beverages may interfere with calcium absorption, due to the phosphoric acid. Caffeinated soft drinks are not a good choice if you are trying to increase your calcium levels. (They are also loaded with simple sugars and therefore help to pack on the pounds due to fat.) Calcium absorption decreases as we age so it's important to start taking calcium supplements or get plenty of calcium in your diet. We also don't make as much vitamin D as our skin ages. By age 80 Americans lose 2–3 inches in height on average due to spongy disks thinning between the vertebrae. We can minimize this loss by getting enough calcium, vitamin D, and exercise.

Hormones help regulate bone remodeling: (remember, the hormones are coming out of endocrine glands.) The body responds to the amount of calcium in the blood. If there is an excess of calcium, the osteoblasts will start to build bone and therefore store calcium for later use. If there is a shortage of calcium in the blood, the osteoclasts will start to break down bone to release the calcium for other bodily functions. Much of this is under hormonal control. The brain regulates calcium levels in the blood and signals the release of the different hormones depending on what is needed.

1. Growth hormone (GH) is released from the pituitary gland (in the brain) and is responsible for the general growth of bones. It also prevents the epiphyses from hardening. The **epiphyses** are the ends of bones, which do not harden during childhood and adolescence so that bones can lengthen. (The epiphyses are mostly cartilage until growth stops and then the cartilage becomes bone). Growth hormone is the primary hormone responsible for the growth spurt adolescents have during puberty.
2. Calcitonin is produced by the thyroid gland (located in the front of the neck). Similar to vitamin D, calcitonin accelerates Ca⁺ absorption. It stimulates osteoblasts to build bones when there is a lot of Ca⁺ in the blood.
3. Parathyroid hormone (PTH) is secreted from the parathyroid gland (located at the back of the neck behind the thyroid gland). This hormone stimulates osteoclasts to break down bone and release Ca⁺. A shortage of Ca⁺ in the blood will trigger this response. (Alcohol also stimulates parathyroid hormone—another reason that alcoholics are more prone to osteoporosis.)
4. Testosterone and estrogen (released from the gonads) also play a role in bone remodeling because they can stimulate osteoblasts (bone building).
5. Cortisol (from the adrenal glands) increases bone breakdown and can decrease bone formatation.

Types of bones include:
- Long bones (leg, arm, finger, toe): All help to absorb stress.
- Short bones (wrist and ankle): Thin compact layer, but the rest is spongy.
- Flat bones (sternum, ribs, scapulae, cranial bones): Thin but protective, irregular (vertebrae and facial bones): very complex shapes.

And there are other bones that don't fit into any of these categories.

The skeleton is divided into the axial (central) and appendicular portions, but I will not cover the names and functions of all of the bones in the body in this book. (One trivia fact that I will mention, however, is that all of the bones are connected to other bones in the skeleton except for the hyoid bone under the tongue, which is suspended there by throat muscles. You never know when you might need that tidbit of information!)

Skeletal Disorders

Osteoporosis is a bone disease that leads to increased risk of bone fractures and breaks due to weak and brittle bones. (Hip and vertebral fractures are the most serious.) Osteoporosis occurs when the bone tissue is broken down by the osteoclasts faster than it is replaced by the osteoblasts during bone remodeling. Osteoporosis affects an estimated 44 million Americans, over 50% of whom are postmenopausal women. Risk factors for osteoporosis are smoking, alcohol abuse, long-term use of certain medications (for asthma, lupus, thyroid problems, and seizures), family history, small frame, low exercise levels, and low calcium intake. Children who don't get enough exercise and calcium are more at risk for osteoporosis later in life because it is important to build healthy bones early on. There is no cure for osteoporosis, but there are medications that can help and a healthy lifestyle can prevent it. Exercise helps to stimulate bone deposits and taking calcium and vitamin D supplements later in life can help prevent this disease.

Bone cancer (osteosarcoma) is curable if caught early; treatment is often amputation of the affected bone.

Broken bones can heal themselves, but the ends of the fractures should be aligned and immobilized during the healing process. (My brother broke his femur straight through and the doctor had to stick a metal rod through the spongy portion in both ends. Eventually the two halves grew back together from bone remodeling. The doctor was never able to get that rod out, however, so his leg sets off the security devices at the airport!)

Our bodies heal themselves relatively well when bones or joints are damaged, but as we age, the healing process slows down. Anything that weakens the immune system, like stress and smoking, can also slow the healing process since the healing begins with inflammation, an immune system response.

There is a three-step process to heal broken bones:
1. Reactive phase: Blood vessels constrict to stop the bleeding in the injured tissue and blood clots form.
2. Reparative phase: Cells grow and transform in the injured area and chondroblasts from into hyaline cartilage.
3. Remodeling phase: Tissue is shaped and replaced, usually making it stronger and harder than it was before the injury.

Joint Problems

Synovial joints can be damaged when ligaments around them are torn (during a sprain) or strained (stretched too far). Sprains in joints (usually in knees, hips, shoulders, and ankles) cause swelling and bleeding from broken blood vessels. (Treat with ice-cold compress the first day and then with a heating pad to increase blood flow to the injured tissue to help speed up the healing process.)

Arthritis is the general term for joint damage and inflammation. There are hundreds of types of arthritis but the most common one is **osteoarthritis**. This refers to a degenerative joint disease that results from either age, infection of the joint tissue, or excessive wear on a joint, often caused by stress from repetitive use. Erosion of the joint tissue leads to the bones rubbing into one another (especially when the protective cartilage between the bones has completely degenerated.) This causes pain and inflammation. Inflammation of a joint can also lead to pinched nerves and tendons associated with the area, leading to more pain. Osteoarthritis is the number-one cause of disability in the U.S., affecting more than 27 million people! Another type of arthritis is **rheumatoid arthritis**, which we already discussed in Chapter 4. This is an autoimmune disorder where the body attacks the joint tissue.

Chapter 8

Circulatory System

The circulatory system is composed of three main parts, the **heart**, **blood vessels**, and **blood**. The heart is the most durable muscle in the body. We have about 60,000 miles' worth of blood vessels of different diameters in our body. Many of these are lined with smooth muscle, constantly contracting, to move the blood around the body. The blood transports oxygen and glucose to our cells for cellular respiration to make energy. It also delivers water, other nutrients, salts, hormones, and helps to take away the cells' wastes. Without the circulatory system providing these essential functions for life, the cells would die. (Think of the skin cells that don't have capillaries to support them on the outer epidermal layer. They all die by the time they reach the surface of the skin. Another example is that brain cells can die within 4–5 minutes of being deprived of oxygen.) Blood circulation is essential for homeostasis (to keep the body's internal chemical and physical conditions within a certain range tolerable for life). Let's go over the importance of the circulatory system in regard to some of the other main systems in our body. Of course, every system needs blood—integument, muscular, skeletal, nervous, reproductive, endocrine, etc.—but here are a handful of the systems that are extremely interrelated with the circulatory system, along with the reasons they're so interrelated.

Five Systems that Need Blood Circulation

1. **Digestive system** (stomach, small intestine, etc.)—the food that we eat is broken down into water, salts, and organic molecules. The digestible particles are then absorbed into the blood (mostly from the small intestines). The blood delivers these molecules to the liver first for processing, and then the particles needed by the cells travel out of the liver for delivery throughout the body. (The indigestible particles are excreted as feces.)

2. **Hepatic portal system** (liver)—the particles from the digestive system are shunted to the liver by the hepatic portal vein (one of the only veins that does not deliver blood directly to the heart). Whereas the Golgi body is the UPS of the cell, the liver has similar functions for the body. It processes, stores,

detoxifies, ships, and modifies molecules depending on what the body needs. Blood enters the liver through both the hepatic arteries (delivering oxygen) and the hepatic portal veins (from gastrointestinal tract and spleen). Once the liver has done its jobs purifying and modifying different molecules, they will go back into the hepatic capillaries and then leave the liver through a different hepatic vein that leads to the heart. The blood now has the necessary nutrients minus the toxins to be delivered to other parts of the body.

3. **Respiratory system** (lungs)—this very important system is where gas exchange takes place. The capillaries in the lungs wrap around the alveoli to pick up oxygen that is inhaled. In the process, CO_2, produced during cellular respiration, diffuses from the blood into the alveoli and then is exhaled from the body. One of the critical functions of the blood is to deliver oxygen, needed by every cell for cellular respiration (otherwise the cells would die), and another is to rid the body of wastes. CO_2 is probably the most prevalent waste in the body.

4. **Urinary system** (kidneys)—filters the blood constantly. The kidneys process the blood about thirty times a day, which means ~45–50 gallons of fluid gets filtered through these amazing organs every day! The kidneys eliminate wastes in the blood (such as ammonia in the form of urea), balance electrolytes, and reabsorb important nutrients, salts, and water that the body needs to keep for delivery to the different cells.

5. **Lymphatic system**—we already mentioned that one of the functions of the lymph is to transport solutes and fluids that leak out of capillaries back to the blood, and the circulatory system helps the lymphatic system by transporting some of the leukocytes and proteins needed for the immune system. The blood also carries platelets, needed for blood clotting, which is also important for the immune system during injury.

So these are some of the critical functions of the circulatory system and how it is connected to the other organ systems of the body. The blood needs to deliver oxygen, nutrients, and ions to all systems. It helps deliver calcium needed for the bone, muscular, and nervous systems as well; think of all the hormones it is delivering all the time for the endocrine and reproductive systems! The blood is critical for maintaining the homeostasis of the body as well by regulating internal temperatures (transporting heat between the body core and skin). It delivers some immune system cells and proteins, forms blood clots to prevent blood loss during injury, etc., etc. I could go on and on about the functions of this system, but let's look now at the parts of the circulatory system, and how they work together to perform these functions. We will also go over some of the disorders of this system, which are usually severe, often fatal, due to the importance of the circulatory system functions. Luckily medical treatment is very effective in helping people with heart, vessel, and blood problems if the problems are caught in time and treatment is available.

Blood Composition

Approximately 55% of blood is **plasma**, which is 90% water, to help carry the other 10% around. The remaining 10% is made up of:

Plasma proteins (make up 7–8% of plasma) include <u>antibodies</u> for the immune system, <u>hemoglobin</u> (a protein containing iron in the center) to carry oxygen, <u>albumin</u> to carry waste, <u>fibrinogen</u> (for blood clotting and osmotic balance), and <u>hormones</u> (some are proteins and some are lipids).

The remaining 2–3% of plasma is composed of glucose, amino acids, lipids, mineral salts, oxygen, and carbon dioxide (mostly in the form of bicarbonate ions).

The other 45% of the blood contains roughly 94% **red blood cells**, 5.8% **platelets**, and a mere 0.2% **white blood cells**.

Red blood cells (erythrocytes) are suspended in the plasma. (A hematocrit measures the proportion of red blood cells [RBCs] in the blood, usually ~38% for women, 48% for men.)

We always have about 20–30 TRILLION red blood cells circulating in our body. (The total number depends on size, sex, health, etc.) Adults have 80–120 trillion cells total, so blood cells make up about a quarter of our cells and RBCs are much smaller than most of our other cell types. They are also unique in the fact that they do not use oxygen for metabolic purposes, and do not have mitochondria, Golgi bodies, endoplasmic reticula, or a nucleus! They use glucose with anaerobic pathways for making energy, and their lack of a nucleus and their biconcave shape provide more surface area for carrying oxygen with the hemoglobin proteins. Red blood cells are mostly water but of the dry ingredients in the cell, 97% is hemoglobin proteins! These transport proteins bind to oxygen with the help of iron. This enables red blood cells to deliver oxygen to every tissue in our body, and not consume any oxygen in the process since they do not use it themselves.

Platelets are irregular fragments of white blood cells that also don't have a nucleus but are important for blood clotting. Platelets survive only about 10 days before they are broken down in the liver and spleen.

There is also a small fraction of white blood cells (leukocytes) in the blood. They squeeze out of blood vessels to fight pathogens in tissues or break down dead or dying cells in a clean-up process.

We have about 5.6 liters of blood in our body. Each day this 5.6 liters goes around the circulatory system so fast that it's like cycling 2,000 gallons of blood through the heart a day.

Each blood cell can cycle through the whole circulatory system (including both pulmonary and systemic systems) within 60 seconds! A blood cell travels at roughly 10 mph if you calculate how far it has to go in all of the vessels to get from the aorta back to the vena cava! So within an hour, the blood has cycled through ~60 times and has traveled 10 miles!

We need a constant supply of blood cells, which are made in the red marrow of spongy bone tissue from stem cells. Red blood cells live ~100–120 days before they lose their organelles and metabolic activities and die. Old RBCs are broken down in the liver and spleen. The kidneys regulate if more erythrocytes are needed by monitoring the levels of oxygen in the blood. If oxygen levels decrease, the kidneys release the hormone erythropoietin (EPO), which then stimulates the production of more RBCs in the bone marrow.

(We can make more than 2 million RBCs per second!) The kidneys can also trigger yellow marrow to convert to red marrow if there is not enough red marrow to produce red blood cells.

Dead blood cells are "eaten" or cleaned up by macrophages (large white blood cells).

We will talk about different blood types (A, B, AB, and O) when we get to Genetics.

Blood clotting:

If a blood vessel is damaged (from punctures or tears) the smooth muscle in the vessel will contract in an automatic response called a spasm. This slows down the blood flow or stops it altogether. Platelets then clump together in the damaged area to create a temporary plug. Platelets store the neurotransmitter serotonin (which is produced mostly in the gut as well as in the brain). The platelets release the serotonin during clotting, which helps to prolong the spasm and attract more platelets to the site. The blood then clots. The whole process usually takes about 2 to 5 minutes. Calcium and vitamin K are needed for the functioning of enzymes involved in the clotting process, as well. If you are deficient in these nutrients, it will take longer to form clots; if you don't have them at all, you can bleed to death. Fibrinogen is a protein that is also needed for clotting. It forms fibrin in a web-like pattern that traps blood at the site of the wound. Sometimes we can see an internal clot when it forms a bruise. When the fibrin web dries and hardens externally, we call it a scab. Phagocytes will then go to work cleaning up the debris at the wound site and eventually the tissue will repair itself. There is a hereditary sex-linked recessive disorder called hemophilia, which inhibits blood clotting. People with this disorder can die from blood loss. Blood clots can form inside blood vessels even when there is no damage to the vessel, and can be very dangerous (can lead to embolisms, strokes, and heart attack). Aspirin can reduce the aggregation of platelets and is sometimes taken to prevent the prevalence of blood clots.

Blood Vessels

1. **Arteries** travel **away** from the heart, and have **oxygenated** blood* (picked up from the lungs). Hemoglobin is the protein that carries O_2 with the help of iron. The complex is called oxyhemoglobin, and the blood in arteries appears red. Arteries, arterioles, venuoles, and veins all have four layers of cells. The outside layer is epithelial (mostly collagen to anchor the vessels to whatever tissue it is running through), inside of that is connective **tissue** (fibrous with elastin, to give flexibility and strength), the next layer is smooth muscle (constantly contracting), and the very inside layer is epithelial tissue again (called endothelial since it is on the inside).
2. **Arterioles** are slightly smaller in diameter than arteries, but continue to carry the oxygenated blood from the arteries to the capillaries.
3. **Capillaries** are only one-cell layer with many pores in this **epithelial** tissue. They have oxygenated blood in the first portion of the capillary bed and deoxygenated blood in the last portion. These very thin vessels are critical, because this is where diffusion of nutrients, gases, and wastes occurs from the blood to the cells and vice versa all over the body. Water and solutes also leak from the capillaries into the interstitial fluid, but as we mentioned in Chapter 4, the lymphatic system helps pick up these items and return them to the circulatory system. Capillaries are so thin that it takes ten capillaries to be as thick

as one strand of hair. Red blood cells travel through capillaries in a single-file line since they are so thin. Capillary beds have precapillary sphincters to regulate the flow of blood into the capillaries. If the tissue doesn't need much blood, the smooth muscles in the sphincter are contracted to tighten the opening, and if the tissue needs more blood flow, the sphincters are relaxed. (During exercise the sphincters are relaxed in capillary beds around the skeletal muscles to help deliver more oxygen to these cells; whereas when you are sleeping and don't need as much blood flowing to your skeletal muscles, the sphincter muscles are contracted here.)

4. **Venuoles** pick up the **deoxygenated** blood from the capillaries to deliver it to the veins.
5. **Veins** deliver the deoxygenated blood* back to the heart. Blood in venuoles and veins appears purplish-blue through the skin, but in the body the blood is still red. It's just a deeper maroon when the hemoglobin is depleted of oxygen, but instead has hydrogen ions attached (then it's called reduced hemoglobin); I will explain this more at the beginning of the respiratory chapter. Veins have valves along the way to prevent blood from backing up (due to gravity). This ensures blood will always be traveling in the right direction. Veins also have a larger lumen (the inside tube) than arteries, whereas arteries have a thicker muscular portion, since arteries are working hard to pump blood everywhere.

*Exception to the artery/vein rule of carrying oxygenated/deoxygenated blood: the pulmonary artery is deoxygenated and the pulmonary vein is oxygenated since it is coming from the lung. (There are other exceptions as well, such as the hepatic portal system, but I'm just simplifying things for this book.)

To give you a sense of how many vessels you have in your body and the proportions, the body has about two miles' worth of arteries, arterioles, veins, and venuoles combined versus 62,000 miles of capillaries! There is one aorta pumping blood out of the heart, 300 arteries and veins, 500,000 arterioles and venuoles, and roughly 40 billion capillaries! Blood flows very rapidly in arteries, slows down as the vessels get smaller in arterioles, and then creeps along through capillaries, allowing time for diffusion, and then speeds up again on its way back to the heart. (In order to keep the same rate from arteries to capillaries, the number of vessels increases so that the blood can branch out into all of the thinner vessels. Think of a river branching into many different streams and then even more little creeks, which then funnel back eventually to another even wider river.)

Heart

The heart is located in the chest cavity a little left of center, between the lungs. It is connected to the lungs by pulmonary veins and arteries on each side for gas exchange. It is the size of a fist and weighs about 300 grams. It is protected by the rib cage and is made up of three layers, including the heart muscles (myocardium, which is sandwiched between the epicardium and endocardium) and surrounded by a tough fibrous sac called the pericardium (a serous membrane). The pericardium protects and lubricates the four heart chambers. The inside of the heart chambers (endocardium) are lined with endothelial tissue (similar to the inside of blood vessels). The four chambers of the heart are divided into the lower portions: right and left ventricles, which pump blood out of the heart; and the upper portions, the right and left atria, which receive blood entering the heart from veins. The <u>septum</u> is a thick wall dividing the two sides of the heart. The chambers are connected by valves, which keep the blood flowing in the right direction through the heart. The "lubb-dupp" you hear when listening to a **heartbeat** is actually these valves closing. The atrio-ventricular valves (tri- and bicuspid [mitral]) close first and the semilunar/aortic valves close soon afterward. The heartbeat is less than a second and is the result of the contraction and relaxation of the chambers. You can also tell how fast your heart is beating by feeling your **pulse** (easiest on the wrist or neck), which is a surge through the arteries every time the ventricles contract. (This indicates how fast the heart is beating.) "Normal" resting pulse rates are usually around 60–80 beats/minute, but they can vary depending on the person's weight, sex, activity levels, health conditions, etc. Rates go up dramatically during exercise in order to deliver oxygen to the cells that need it quickly. The heart beats around 100,000 times a day, which results in about 2.5 billion beats in a 70-year life! What an amazing organ the heart is!

Blood pressure (systolic/diastolic) is the pressure of the blood against the vessel walls and is measured during two phases of the cardiac cycle. The **systolic pressure** is the highest pressure, and refers to the force of blood in the arteries when the ventricles contract. (When the ventricles contract to force blood into the pulmonary artery and aorta, the atrioventricular valves close so that blood won't back up into the atria—this is the "lubb" you hear from a heartbeat.) Then when the ventricles relax, the aortic and semilunar valves close to prevent the blood from going back into the ventricles ("dupp"). The **diastolic pressure** refers to the force of blood on the walls of the circulatory system while the heart is in complete diastole (relaxation). The pressure during relaxation (rests between beats) is when the right and left atria fill with blood from the veins and then move the blood into the ventricles. Blood pressure is measured in millimeters of mercury (mmHg). A healthy blood pressure is **less than 120/80 but more than 100/60**. Above this range is considered pre-hypertension or hypertension (once the numbers are up to 140/90 mmHg or higher), and numbers below the normal range can indicate hypotension.

Hypertension (high blood pressure) is often called the "silent killer" since it rarely shows symptoms, but can lead to ruptured blood vessels, stroke, heart disease, kidney damage, and memory or vision problems if left untreated. Twenty-five percent of Americans have hypertension. It can be remedied by exercise and maintaining a healthy weight, cutting back on salts, high-cholesterol foods, and alcohol, and by not smoking, and limiting stress. (Stress triggers the release of stress hormones and neurotransmitters, which can raise heart rates and blood pressure, both of which can weaken the heart over time.) If diet and lifestyle changes don't help, there are also hypertension medications. (Some symptoms of severe hypertension can be nosebleeds, lightheadedness, dizziness, nausea, vomiting, chest pains, and headache, but generally the only way to find out if you have hypertension is to get your blood pressure checked periodically.)

Hypotension (low blood pressure) is not usually as dangerous, but can indicate that there is not enough water or proteins in your plasma, that you're not making enough blood cells, or that too much blood was lost during injury. If blood pressure is extremely low, not enough oxygen and sugar will be delivered to the brain. Symptoms include dizziness, fainting, weakness, blurred vision, and sleepiness. It can be treated by drinking plenty of fluids and increasing salt intake (if it's not a blood-cell issue), injection of more blood, or medication. Alcohol and some medications such as anti-anxiety, anti-depressants, diuretics, and painkillers can all cause hypotension. Hypotension can also just be hereditary and not cause any symptoms. (My blood pressure is usually around 95/55 and I seem to be okay.)

Now let's look at the **flow of blood** through the different chambers, valves, arteries, and veins of the heart and how it is connected to the lungs. There are two circulatory circuits: the pulmonary circuit is where the blood picks up oxygen from the lungs and the systemic circuit is where blood travels to and from all of the other tissues of the body. Let's start from where the deoxygenated blood enters the heart on the right side. The way I will describe the heart makes it sound like the blood is going in succession from one chamber to the next in a linear order, but in reality the two ventricles are pumping blood to the arteries (pulmonary and aorta) at the same time, the two atria are receiving blood from veins at the same time, and ventricles are receiving blood from the atria simultaneously. Think of the heartbeat cycle and know that during the first sound both ventricles are contracting, and after the second sound both ventricles are filling with blood again from both atria. I know this sounds confusing but hopefully once we go over it in class it will make more sense.

There are four chambers of the heart enclosed by thick muscular walls.

Deoxygenated blood from the head, arms, shoulders, and upper chest region enters the heart through the **superior vena cava** into the **right atrium** of the heart.

Blood from anywhere lower than the heart is also entering the right atrium, but it comes through the **inferior vena cava** (both venae cavae are very large veins).

From the right atrium the blood must pass through the **tricuspid** valve to get from the right atrium to the right ventricle (think of tricuspid as "trying" to get oxygen, since this is deoxygenated blood). This valve is also called an atrioventricular valve since it separates an atrium and ventricle.

The blood goes from the right ventricle to the **pulmonary artery**, which branches to pick up oxygen from the lungs on each side of the heart. (Gas exchange occurs by diffusion through the right and left capillary beds in the lungs.) Arteries go away from the heart but usually carry oxygenated blood; remember, this is an exception to that rule: this artery is deoxygenated. There is also a valve between the right ventricle and pulmonary artery called the **pulmonic valve** (or semilunar valve). This valve keeps the blood from going back into the right ventricle. The oxygenated blood from the capillaries in the lungs travels to right and left **pulmonary veins**, which bring the blood back into the heart to the left chamber and into the **left atrium**.

From the left atrium the blood passes through the **bicuspid valve** (also called the mitral valve or AV valve—think of "bye-bye" as the blood is now on its way out of the heart) to the **left ventricle**.

The left ventricle is the largest, most muscular compartment of the heart since it needs to pump the blood out of the heart through the **aortic valve** (another semilunar valve) and then into the **aorta**—the largest vessel in the body.

This will lead to numerous arteries (some going up, others going down) that branch into arterioles and then into capillary beds, which are located all over the body (including on the heart itself) to bring nutrients and oxygen everywhere.

After depositing oxygen and nutrients to cells and picking up carbon dioxide and wastes, the blood travels back to the right side of the heart from capillaries to venuoles to veins to the superior and inferior venae cavae to start the process all over again.

The **sinoatrial (SA) node** (also called the **pacemaker**) located on the wall of the right atrium sets the rate of the heartbeat by sending out an electrical signal to start the contraction of the heart muscle, which then causes a domino effect and the rest of the heart contracts in the same rhythm. Cardiac muscles are the only muscles that can produce rhythmic contractions without nerve impulses. In this way the SA node governs the resting heart rate. The atrioventricular node, at the base of the atrium, delays the contraction signal before sending it onto the ventricle.

The heart rate can also be governed by the cardiac center in the brain (the medulla oblongata). With the help of different hormones and neurotransmitters, the heart rate can increase or decrease depending on the body's needs. These signals can override the pacemaker (intrinsic) heartbeat.

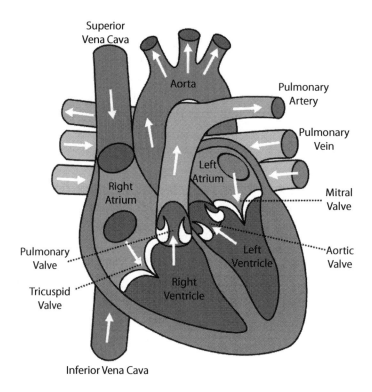

Here is a recap of the order in which blood goes through the heart: **the italicized portions signify the pulmonary circuit and the rest is the systemic circuit**. Remember that the blood is being contracted by the right and left ventricles at the same time to force the blood into the pulmonary and systemic circuits simultaneously. In the flow chart below, the two systems are working simultaneously and are separated by my "reminder." I've indicated where the "lubb-dupp" of the heartbeat occurs when the valves close after the blood passes through them. As the ventricles begin to contract, the semilunar valves (pulmonic and aortic) are forced open to allow blood through the open valves into the pulmonary artery toward the lungs, and into the aorta, toward the body.

When the ventricles finish contracting and begin to relax, the aortic and pulmonic valves snap shut. These valves prevent blood from flowing back into the ventricles. This pattern is repeated with every heartbeat, causing blood to flow continuously to the heart, lungs, and body.

(Deoxygenated blood) Vena cava → right atrium → tricuspid ("lubb") → right ventricle contracts → pulmonic (semilunar) valve ("dupp") → *pulmonary artery → arterioles → right and left pulmonary capillary beds* →

Reminder: This next part is happening simultaneously (the "lubb" and "dupp" match up since the bicuspid and tricuspid close at the same time and the pulmonic and aortic valves close at the same time.):

(Oxygenated blood) *venuoles → right and left pulmonary veins* → left atrium → bicuspid ("lubb") → left ventricle → aortic valve ("dupp") → aorta → arteries → arterioles → capillaries → venuoles → veins (and then back to vena cava)

> *In biology there seem to be many names for the same things—just to clarify, there are two "semilunar" valves (which do not have chordae tendineae) leading to the pulmonary artery and the aorta, and these are the pulmonic and aortic valves. Then there are the valves between the atria and ventricles called atrioventricular valves (AV) that <u>do</u> have **chordae tendineae** ("heartstrings"), which are tendons made of cartilage and elastin connecting the papillary muscles of the heart to the valves. These valves are the tricuspid (three flaps) and bicuspid (two flaps) valves, but just to make it more confusing, the bicuspid valve is also called the mitral valve. Sometimes words are spelled differently too: "venuoles" is the same thing as "venules" (I like "venuoles" better).

All blood cycles through the pulmonary circuit, but as for the systemic circuit, there are different percentages of blood that go to different areas depending on the body's needs. The highest percentage of blood goes to the digestive (21%), excretory (20%), and skeletomuscular systems (15%), and the brain (13%).

Cardiovascular Diseases

The number-one killer in the United States is heart disease. Sixty-four million Americans have some sort of cardiac problem, either <u>congenital</u> (born with the disorder) or <u>acquired</u> (heart problems can develop anywhere from childhood on). Congenital disorders are not always from genetic disorders; they can just be random malformations of the heart during fetal development. These congenital disorders can often be corrected by surgery soon after birth. (Eight out of a thousand infants have congenital heart defects!)

Some different types of heart disease and causes:

Heart Failure is the most expensive health problem in the United States. It is a general term used for when the heart is not pumping blood (and therefore oxygen and nutrients) as efficiently as it should. This can be due to a number of problems that I will discuss here, such as clogged or weakened vessels, valve defects, etc.

Coronary Artery Disease Caused by Atherosclerosis (or Arteriosclerosis) ("hardened vessels") is the number-one heart disease in adults. This is when arteries are clogged or weakened by the accumulation of plaques composed of fats, cholesterol, dead cells, macrophages, and calcium. This leads to hardening of the arteries (or arterioles) and can either decrease the blood flow through the vessels to an area or cause hypertension by creating a small diameter in the vessels that need to transport a lot of blood. If too much pressure builds up, a vessel can burst. Even just a 25% decrease in blood delivery to the heart can lead to angina (chest pain) and/or a heart attack. (A **heart attack** is when not enough oxygen and nutrients are delivered to the cardiac muscle cells and the cells die. If enough of these cells die, the heart stops pumping blood completely. Fifty percent of people who have a heart attack die within an hour if not treated quickly.) The primary reason for atherosclerosis is high LDL cholesterol levels*. Cholesterol sticks to the endothelial cells in the vessels and attracts blood clots to the site, which then form even more of a barrier around the plaque. This can be lethal if they block the blood flow completely, if the vessel bursts, or if the clot breaks free and plugs a smaller vessel. A blood clot in place is called a **thrombus**; if it moves to another area it is called an **embolism**. An embolism in the brain can lead to a **stroke**. (See below.)

> *There is good cholesterol (high-density lipoproteins, HDL) that takes away the cholesterol and triglycerides from the vessels and delivers it to the liver for processing and eventual excretion, and there is bad cholesterol (low-density lipoproteins, LDL) that carries cholesterol in the blood and causes clogged vessels. We will talk more about these lipoproteins in Chapter 11 (Nutrition).

Hypercholesterolemia refers to high cholesterol in the blood (high LDL levels) and can lead to heart disease and stroke. (LDL levels should be under 129 mg/dL, and total LDL + HDL levels should be under 239 mg/dL.) HDL levels are best above 60 mg/dL, since HDL helps remove cholesterol from blood.

Blood clots (thrombus) from platelet accumulation block blood vessels and can stop the flow of blood altogether. People with narrowed arteries due to plaques (atherosclerosis) are even more at risk when platelets clot. If the blood can't reach the heart muscles (through the capillary beds surrounding the heart), the heart will stop contracting due to lack of oxygen unless blood flow is restored within minutes. Blood clots that break free from the site of origin to another site to cause blood vessel blockage are called **embolisms**. Embolisms often lodge in the brain (can cause a stroke) or in the lungs.

Pulmonary embolism kills roughly 60,000 people per year but occurs in over 200,000 people, so it is not always fatal. If you have to sit for a long period of time, be sure to get up and walk around every 2 hours to get your circulation flowing normally and don't cross your legs for extended periods since pulmonary embolisms often start out as blood clots in the legs when the circulation is decreased.

Stroke is a term used to signal the loss of brain function due to lack of blood flow to the brain (resulting in insufficient oxygen and sugar). (Brain cells start to die within 4–5 minutes of not receiving enough oxygen.) A stroke can occur if a blood clot moves into the brain (embolism) or forms there, blocking the flow of oxygen to the neurons and glial cells in the brain. It can also occur if there is a leakage of blood in the brain from a hemorrhage, such as if a blood vessel bursts due to an aneurysm. Strokes can be severe (causing death) or non-fatal (usually causing loss of some sort of motor coordination, sometimes paralysis, loss of sensory perception, speech, or cognition—it depends on where the damage occurs and how long the brain is deprived of oxygen). In the United States approximately 700,000 people every year suffer a stroke. If a person suffers a stroke s/he needs to get to the hospital **as soon as possible** (definitely within the first few hours) in order to possibly save brain function or their life. Here are important things to check if you think someone has suffered a stroke:

S: Smile and be able to stick the tongue out straight

T: Talk—be able to say a simple sentence

R: Raise both arms

If the person cannot perform one or more of these tasks, get them to the hospital immediately.

Heart attack is called a **myocardial** (heart muscle) **infarction** (tissue death or necrosis due to lack of oxygen). Heart attacks usually occur when plaques or clots in a coronary artery stop the blood flow to the heart muscles. Risk factors include smoking, obesity, hypertension, blood clots, and clogged arteries due to high cholesterol. Heart attacks are most common around the holidays due to stress and food as well! A Thanksgiving meal can be roughly 4,000 calories and large meals increase insulin and triglyceride levels and put more stress on the vessels and heart. All of this combined can quadruple the chances of a heart attack for people who have other risk factors as well.

> **Heart Attack Symptoms** include pain down the arm, behind the breastbone, or in the jaw area (especially for women); sweating; neck and back pain; nausea; indigestion; shortness of breath; fatigue; and irregular heartbeat. One in five deaths in the United States are from heart attacks! Of the nonfatal heart attacks, 40% of the survivors will die within a year after the myocardial infarction. If you ever think you are having a heart attack, call 911 and chew an aspirin to thin your blood. (Chewing the aspirin will deliver the medicine faster than swallowing it whole.)

Aneurysm—when an artery vessel wall balloons outward due to weakening of the blood vessel. This can be due to an inborn defect, disease, or injury to the tissue. The weakened vessel can stimulate clots to form in the area or they can burst altogether. Aneurysms tend to occur within the aorta; in the arteries near the spleen or intestine; behind the knee; or below the brain. If an aneurysm bursts, it can cause a severe hemorrhage and is often fatal. Cerebral aneurysms usually cause either a stroke or death.

Cardiomyopathy—chronic disease that causes the heart muscles to weaken (usually the ventricle muscles first). Once it spreads all over, the heart can eventually stop beating. This disease is the number-one reason children receive heart transplants.

Arrhythmia—an irregular heartbeat, whether abnormally fast (tachycardia), abnormally slow (bradycardia), skipping beats altogether, or haphazard beats (ventricular fibrillations). They can be from congenital disorders or can be acquired at any time during life for different reasons. (Sometimes triggered by hormonal changes or diet—caffeine and alcohol can sometimes exacerbate the condition.) Some are mild and sporadic and don't need treatment, and some require medication, surgery, or a pacemaker to keep the heart beating regularly.

Valve problems—valves can become hardened (stenosis) and not function properly (leading to heart failure), or can be weak and not close tightly so that blood leaks backward (regurgitation). These valve issues can be congenital or acquired, and there are certain bacterial infections that target valves as well (see endocarditis).

Aortic stenosis—when the aortic valve fails to open fully.

Aortic insufficiency/regurgitation—when the valve is incompetent and blood flows back into the left ventricle.

Rheumatic heart disease—occurs from complications of an untreated strep throat infection (from streptococcus bacteria), and can lead to permanent heart damage and even death. It begins when antibodies the body produces to fight the strep infection begin to attack other parts of the body, such as the heart valves, causing the heart valves to thicken and scar. Inflammation and weakening of the heart muscle may also occur. This is also considered an autoimmune disease. (Be sure to always treat a strep throat infection early with antibiotics!)

Endocarditis—a bacterial infection that travels in the blood to the heart and attacks the heart valves. Sometimes microbes enter the bloodstream from equipment during dental treatment, or more often from needles during intravenous drug use. In either case it is a very serious infection, which can lead to sudden heart failure.

Lyme disease—caused by bacteria spread by ticks—affects many parts of the body, one of which is the heart.

Measles—from the rubella virus—can damage the heart muscle.

Alcohol and Drugs—can lead to weakening of the heart muscle (heart failure) or heart attack (especially with overdoses).

Hypo- and Hypertension—(already discussed above).

Heart cancer—rarely occurs, but sometimes cancer will travel to the heart from other locations.

Varicose Veins—from weakened valves in veins (usually in the legs) that cause the blood to accumulate in certain locations. The veins can stretch and bulge, which can be seen through the skin. It is more common in women due to the fact that female hormones can weaken valves and veins (especially during pregnancy). Exercise can help keep veins healthy and strong.

Blood Disorders

Carbon monoxide poisoning—results from inhaling carbon monoxide (CO) from incomplete burning of gasoline, wood, coal, or paper. It is poisonous due to the fact that when there is carbon monoxide in the blood, hemoglobin will bind with the carbon monoxide two hundred times stronger than with oxygen, and therefore oxygen will not be delivered to cells. This can lead to cell and tissue death, which can be fatal. Carbon monoxide poisoning is one of the most common causes of poisoning in the United States. Over 50,000 people visit the emergency room every year due to CO poisoning and roughly 5,000 people a year die from carbon monoxide. We usually have a certain amount of carbon monoxide in our blood from trace amounts that are generally present in everyday life. Non-smokers have about 1–3% CO in blood. Smokers, however, have more since they are inhaling it with the rest of the toxins in cigarettes. They have roughly 10–12% carbon monoxide in their blood. Once you reach levels of about 15%, CO can cause headaches, nausea, and vomiting. Higher levels of around 20–25% (usually from gas leaks or car exhaust), can cause brain and heart problems and can lead to death once the tissue has been starved of oxygen long enough. Never leave a car idling in a garage, even if the door is open!

Mononucleosis—Epstein-Barr virus causes overproduction of lymphocytes.

Leukemia—results from cancer of bone marrow or of blood, (usually leukocytes/WBCs). The abnormal cancerous cells will destroy healthy blood and bone marrow, affecting the immune system and causing fever, weight loss, pain, and anemia.

HIV—we already went over this virus, but remember that it affects white blood cells.

Septicemia—some bacterial infections destroy red blood cells and prevent clotting.

Anemia, "no blood"—refers to many different types of blood disorders where the levels of red blood cells are insufficient or they are not delivering sufficient oxygen to cells. There are several causes for anemia:

Iron deficiencies: Since iron is needed for the production of hemoglobin, which carries oxygen. (This can be easily remedied by eating more iron-rich foods.)

Folic acid and vitamin B deficiencies: Both of these are needed to produce red blood cells in the bone marrow.

Sickle cell anemia: An autosomal recessive genetic disorder that causes red blood cells to die before their time. They become sickle-shaped and hemolysis (bursting of blood) occurs.

Malaria: Caused by a protist transmitted by mosquitoes. Once it gets inside blood cells, it causes the cells to burst. (Another hemolytic disease.)

For a healthy circulatory system follow these steps:

Eat a healthy diet with plenty of fruits and vegetables (high-fiber diets will help lower your blood pressure), as well as some lean proteins, and whole grains.

Limit saturated fats; these increase LDL cholesterol levels in the blood and can lead to high blood pressure, arteriosclerosis, stroke, or heart attack.

Limit sugar and salt consumption. Too much sugar can lead to obesity and type 2 diabetes, (both are risk factors for heart disease). Too much salt will cause you to retain more water, which puts excess pressure on your blood vessels, leading to hypertension, another risk factor for heart disease.

Alcohol in moderation may actually lower the risk of heart disease for some people, however excessive alcohol use can weaken the heart, damage the liver and brain, and increase the risk of cancer. For women, the recommendation is generally no more than 5–7 drinks per week and the number is slightly higher for men who have a larger body mass.

Keep blood pressure under 120/80 and cholesterol under 239 mg/dL.

Get plenty of exercise. 75–150 minutes per week is recommended for a healthy circulatory system: it helps your heart and blood vessels.

Keep a healthy BMI (Body Mass Index). Obesity is another main cause of heart disease.

Reduce stress levels and get enough sleep.

Don't smoke cigarettes. Smokers are twice as prone to heart disease!

Don't do drugs. Certain drugs speed up the heart and can cause heart attack, stroke, heart valve problems, and other heart diseases.

CPR, cardiopulmonary resuscitation (rescue)—performed when someone's heart or breathing has stopped (cardiac or respiratory arrest). I highly recommend everyone get CPR-certified. It involves delivering oxygen* to the person by blowing into the mouth (while the nose is pinched shut so that the oxygen will travel down to the lungs). If there are still no signs of breathing, you follow with thirty chest compressions to keep the heart pumping. (For the timing of the chest compressions, sing the "Ah Ah Ah Ah Staying Alive" Bee Gees song!) You need to continue this pattern (two breaths, thirty compressions) until help arrives (someone with a defibrillator) or until the person starts to move again. Proper technique is critical, so it's better to be CPR-certified than just trying to figure it out from a video, the 911 dispatcher, or the Internet. But CPR will not kill anyone, so any help is better than no help. (Always call 911 first unless you are helping a baby or a child—then do the breathing immediately, since chances are that the cardiac arrest is due to an airway problem. If you still don't see breathing within two minutes, call 911.) If you do CPR on a person whose heart has stopped beating, there is a 30% chance the person will live if a defibrillator can arrive within several minutes to shock the heart. *Hands-only CPR (no mouth-to-mouth) is recommended for bystanders who don't know the victim or CPR very well:

> The American Heart Association in an Advisory Statement of March 2008 recommends hands-only CPR in the following circumstances: If a bystander, not trained in standard CPR, sees an adult suddenly collapse, then he or she should call 911 and provide chest compressions by pushing hard and fast in the center of the chest. Interruptions should be kept to a minimum until trained rescuers arrive. For bystanders previously trained in standard CPR, hands-only CPR may be performed if the bystander is not confident or is unwilling to perform mouth-to-mouth ventilation. The advisory goes on to state that the recommendation of hands-only CPR is limited to bystanders who directly witness out-of-hospital cardiac arrest of likely cardiac origin (sudden collapse after signs consistent with a myocardial infarction).

In fact, a study in 2007 found that of 4,000 cases of CPR performed on cardiac arrest victims, the people who were treated with just chest compressions and no mouth-to-mouth were twice as likely to survive without brain damage, so the advice from that study was to focus on the chest compressions.

*We always refer to the fact that we exhale CO_2, but exhalation also contains ~16% O_2, which is just a little less O_2 than we get from air when we are breathing regularly. Air has 21% O_2, so yes, by exhaling into a person's mouth you are delivering enough O_2 to resuscitate them. This leads us to the next very important organ system in the body … the respiratory system.

Chapter 9

Respiratory System

S ince this chapter is about the pulmonary system and the number-one threat to the pulmonary system comes from cigarette smoke, I want to start off by explaining some of the reasons that tobacco is the deadliest plant in the world. **Cigarettes are the number-one cause of preventable disease and death**, leading to roughly half a million deaths in the United States every year (and billions in health care expenses—in fact 9 billion a year just in California!). I'm still amazed at how many people smoke despite the general knowledge that cigarettes can kill you. The surgeon general's report on cigarettes came out in 1966 with the warning "Caution: cigarette smoking may be hazardous to your health." This did not deter many people from smoking but hopefully the new warnings that are proposed to come out in September 2012 will have more of an impact. Cigarettes will be labeled with warnings "Smoking can kill you" or "Cigarettes cause cancer," along with graphic pictures on the packaging of people with air tubes in their throats and of damaged lungs. (The tobacco industry is currently fighting against these proposed warnings, of course, so we will see what happens.) I do realize how addictive cigarettes are once you've started smoking (my brother has been trying to quit for years)—I just wish no one would start. If you are a smoker, I hope that reading about the harmful effects of tobacco will inspire you to quit. If you are not a smoker but know one (or many), I hope you will pass along this information as well. Around 50% of all regular smokers will die from cigarettes eventually and half of smokers will die in middle age. This is not a gamble anyone should take!

Harmful Effects of Smoking Tobacco

Smoking impairs the **immune system**. Smoke helps kill white blood cells, and stimulates more mucus, which also kills phagocytes. Smoke immobilizes cilia in the respiratory tract, which function to sweep pollutants and microbes away as well as helps to expel excess mucus. One cigarette can immobilize cilia for hours!

Smokers are more prone to <u>colds</u>, <u>asthma</u>, <u>bronchitis</u>, <u>emphysema</u>, and other <u>respiratory infections</u>. It takes smokers longer to heal from sprains, fractures, and broken bones as well, since the immune system helps with the inflammation and healing process.

Tobacco is a **carcinogen**, which means a substance that can cause **cancer**. Since tobacco also weakens the immune system (which is our first line of defense against cancer), smoking poses a double risk for cancer. Actually, it's more than double since there are more than 4,000 other chemicals in cigarettes and 40 to 60 of them are also carcinogens, including radon, nitrosamine, and benzopyrene. Many of the compounds in tobacco smoke attack epithelial cells and allow them to grow out of control. (And yes, "organic" cigarettes cause cancer as well, and smoking through a hookah pipe may be even worse than smoking cigarettes.) Tobacco causes lung and breast cancers the most, but smoking is also linked to leukemia and cancer of the head, neck, throat, mouth, esophagus, bladder, stomach, kidney, pancreas, colon, and cervix.

Cigarettes cause about 90% of lung cancer deaths per year—roughly 174,000 people! The number of years smoking increases the risk of cancer more than the number of cigarettes smoked per day.

Pipe and cigar smokers have an increased risk for lung and liver cancers as well as stomach, pancreas, colon, and bladder cancers.

Chewing tobacco also causes cancer—the most common types are oral, pancreatic, and esophageal cancers. (Not to mention the bad breath, yellow teeth, problems with the digestive system, etc.)

Smoking increases your chances of **heart disease** and **stroke** because it raises blood pressure, increases LDL cholesterol levels, decreases the HDL cholesterol levels, causes more blood clots, and weakens the heart and blood vessels in general. A person who smokes a pack of cigarettes a day is twice as likely to suffer a heart attack as a non-smoker.

Nicotine constricts blood vessels and can lead to poor circulation and thermoregulation. (Smokers are often cold.)

Nicotine can also suppress appetite in some people, leading to insufficient nutrient consumption.

Smoking can shorten a person's life span by 8.3 years, on average. Every cigarette you smoke takes 11 minutes away from your life, according to researchers in England. If you quit between 35 and 39 you will add roughly 6 to 9 years onto your shortened life span versus if you quit between 65 and 69. This only adds an extra 1 to 4 years to your life, so don't wait. If you're a smoker—quit today! (I sound like an infomercial!)

Cigarettes can alter sex hormone effectiveness, especially in females. Women who smoke have higher **reproductive problems**. Women who smoke and are using oral contraceptives have even more problems in general.

Smoking during pregnancy can cause preterm labor, placental abruption or previa, stillbirth, low birth weights, and sudden infant death syndrome (SIDS).

Smoking can kill nerve cells, as well as sensory receptors, interfering with smell and taste. Smoking also increases the risk of memory loss later in life.

Cigarettes cause increased carbon monoxide and carbon dioxide levels (both of which at higher levels are toxic) and decreases oxygen levels. Not enough oxygen in the blood starves tissues, which leads to health problems and also reduces the ability to exercise (think back on all the benefits of exercise).

Smoking increases the risk of **diabetes**.

Smoking causes skin problems, deep wrinkling, stained teeth, and bad smell!

Secondhand smoke is extremely harmful as well. It causes many types of respiratory illnesses (such as asthma), skin rashes, heart disease, and cancer. Three thousand people per year die from secondhand smoke! (So if you don't think of yourself, try to think of those around you when you are smoking—especially don't smoke near babies and children since they have more sensitive lungs.)

The **pollution** resulting from cigarette smokers (butts left everywhere) costs millions of dollars in clean-up efforts—cigarette butts are the number-one littered item in the world! They also affect other organisms that try to ingest them. San Francisco now has a cigarette tax to help defray the costs of clean-up. Now we just need a tax on the tobacco industry to pay for the billions of dollars in healthcare costs cigarettes cause every year.

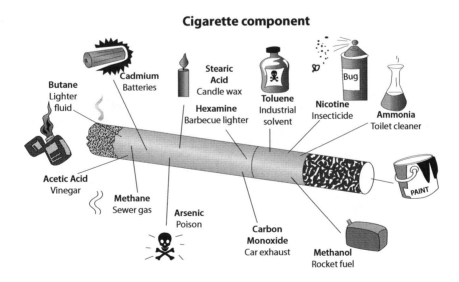

Cigarette component

Congress wants the FDA to regulate cigarettes due to the huge health problems that cigarettes cause, but it's hard to get around the 70-billion-dollar tobacco industry! The FDA has already banned candy-flavored cigarettes (targeted at kids), and banning menthol would be helpful, but it probably won't happen. (The menthol flavoring masks the harsh taste of cigarettes and is preferred by young smokers and African Americans, who have the highest rate of smoking-related diseases.)

Fifty percent of long-term smokers eventually die of smoking-related causes, such as heart disease (number-one killer in America), lung cancer (number-one cause of death by cancer in both men and women), stroke, infections, respiratory ailments, etc.

So the bottom line is, if you are a smoker, PLEASE STOP SMOKING, and if you're not, DON'T START! Also limit your exposure to secondhand smoke, since this is a killer as well. Some people start smoking as teenagers since they think it is "cool." When my 7-year-old daughter heard this she said, "Why don't they just go to the beach?" She's right, going to the beach is much "cooler" and a lot healthier. If the health concerns don't convince you to stop smoking, think of how much money you can save by not smoking. If you smoke a half a pack per day, this can end up costing you over $1,000 per year! (There is an app called "quitter" available for free from iTunes that calculates how much money you save by not buying cigarettes.) And if you STILL want to smoke, at least throw your butts in the trash and smoke away from other people. Okay, I'll get off my soapbox now and get on with the respiratory system.

Respiratory System Main Function: Gas Exchange

We breathe in air, which is a gas, composed of 21% oxygen, 78% nitrogen, and traces of water vapor, carbon dioxide, helium, argon, and various other components. The molecule that we need for cellular respiration is oxygen, which binds with hemoglobin to form oxyhemoglobin in our blood. Then we breathe out the remains of the oxygen (not all oxygen is delivered to the lungs and blood; some comes back out with our breath) as well as carbon dioxide that the body produces during cellular respiration. Bringing in oxygen and expelling carbon dioxide also regulates the pH levels of the blood, a very important part of homeostasis.

Carbon dioxide travels in the body in different forms:

Seven percent stays in the blood plasma

Ninety-three percent exits the body through exhalations. This CO_2 travels in the blood in different ways:

Twenty-three percent binds with hemoglobin

Seventy percent combines with H_2O inside the blood to form bicarbonate. Most of the carbon dioxide travels in the blood in this form and then it gets converted back to its original form before it is expelled. (This is why I have arrows that go in both directions in this equation—it is a reversible reaction.)

$$(CO_2 + H_2O \Leftrightarrow H_2CO_3 \Leftrightarrow HCO_3^- + H^+)$$

H_2CO_3 is carbonic acid, which dissociates quickly with the help of the enzyme carbonic anhydrase to form HCO_3^- (bicarbonate) and H^+ (hydrogen ions). The acidic hydrogen ions that are produced in this reaction attach with hemoglobin in the blood (called "reduced hemoglobin"). Once it travels to the lungs, the reaction reverses, and the hydrogen ions detach from the hemoglobin to recombine with the bicarbonate to form CO_2 and water. Both the CO_2 and small amounts of H_2O will exit with every exhalation. Meanwhile the free hemoglobin is now ready to pick up oxygen coming from the inhalation into the lungs. CO_2 is a waste that needs to be expelled, but it also has an important function in the blood. When it is converted to bicarbonate, it serves as a **buffer** in the blood. This is important, since the other product of this reaction, hydrogen ions, raises the acidity of the blood. Without this buffer we would not be able to monitor the internal pH of the blood and we would die. (If our blood was too acidic, the acids would dissolve cell membranes and the contents would spill out, thus killing the cells.)

The medulla oblongata and pons in the brain stem can detect the acidity of the blood (through chemo-receptors in the carotid artery and aorta). If the blood is too acidic (high concentration of hydrogen ions), the brain will trigger the phrenic and thoracic nerves to contract and relax the diaphragm and inter-costal skeletal muscles that help with ventilating the lungs. This causes us to expel more CO_2 by increasing the depth and rate of respiration. Usually people can't hold their breath longer than a couple of minutes because the brain detects the increased acidity (lower pH) and forces exhalation. When CO_2 levels are stable, the respiratory center of the brain causes rhythmic contractions of the diaphragm: two seconds contracted followed by three seconds of relaxation.

Another interesting fact associated with CO_2 in your blood is that it affects the color of the blood. The H^+ picked up by hemoglobin in the blood ("reduced hemoglobin") causes the blood to appear deep crimson/purple, which looks blue under the skin (our veins look blue because they have deoxygenated blood with plenty of H^+ and CO_2 as bicarbonate). Blood in our arteries does not have CO_2 and therefore doesn't have the reduced hemoglobin (instead the hemoglobin is carrying O_2) and is a lighter red that we usually can't detect through the skin.

The **Respiratory System** is composed of the **nose, airways,** and **lungs**. Here is my crude drawing of the respiratory system so that you can picture where everything is as we are going through the functions of the different parts of this amazing system:

1. Nose
2. Pharynx
3. Larynx
4. Trachea
5. Bronchi
6. Lungs
7. Bronchioles
8. Alveoli

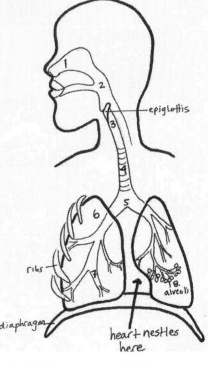

1. **Nose**. When we breathe at rest, the air moves into the nose first (we can also breathe through our mouths, but this is more common when doing strenuous exercise than when breathing at rest). There are many functions of the nose—not all functions are respiratory-related, but I'll list them all here anyway.

   • **Hair** and **cilia** in nose function to filter out large particles, like dust and pollen.
   • Large blood supply in nasal tissue warms the epithelial tissue of nose, which helps to warm and moisten air.

- **Secretion of mucus** by the mucus membranes in the nose adds another layer of defense to deter pathogens.
- Two nasal chambers are separated by the <u>septum</u> (bone and cartilage) with openings at the back and top of nose. The opening between the nose and throat is normally open, but it is closed when we swallow by a flap called the <u>uvula</u>. This blocks liquids from entering the nose from the mouth (when you laugh while drinking, sometimes the uvula spasms and your drink ends up coming out your nose).
- **Tear glands** above the nose produce moisture that drains into the nasal cavity (when we cry, we produce more tears, which mix with the mucus to produce a runny nose).
- The nose also has chemosensory neurons in the nasal epithelial for smell.

2. **Pharynx** leads to both the respiratory system (through the larynx) as well as the digestive system (through the esophagus). The pharynx includes the areas where the uvula hangs down, where the Eustachian tubes connect the nasopharynx to the middle ear, and is where the tonsils are located. The base of the pharynx (laryngopharynx) has two openings going to the larynx in the front and the esophagus toward the back.

3. **Larynx** contains nine pieces of cartilage, one of which is the **epiglottis** at the entrance of the larynx. This flap is attached to tongue muscles and is usually open to allow air into the larynx. However, when we swallow, the tongue pushes against the roof of the mouth, which causes the epiglottis to rest on top of the larynx, bending closed to prevent food or liquids from entering the airways. The food is directed down to the esophagus (to the digestive tract) instead of entering the larynx. (The hyoid bone is located here as well, to provide an attachment site for muscles at the floor of the mouth and tongue.) The larynx also contains the **Adam's apple** (another piece of cartilage). Testosterone stimulates cartilage growth, so men have a more pronounced Adam's apple than women have. The **voice box** is also located in the larynx. Vocal folds (or "cords") are flaps of cartilage held in place by elastic ligaments across the glottis, which enable us to produce sound. We produce sound (talking, singing, etc.) while we exhale. Our exhalation is causing tension against the folds, and the more tension we cause, the higher the pitch. We can modify the sound by manipulating teeth, tongue, soft palate, mouth, etc. Testosterone also affects the cartilage in the vocal folds, causing boys' voices to change as testosterone levels increase during puberty. The larynx opens more, tilts forward, and the cartilage widens, causing a lower pitch.

4. **Trachea**, or the "windpipe." The opening of the trachea is called the **glottis**. The trachea is a tube with muscular walls embedded with hyaline cartilage to keep it open and add strength. You can feel your "tracheal rings" through your skin. The ring doesn't go all the way around, since there is an opening at the back of the throat to allow for the esophagus (which is behind the trachea in the neck) to expand during swallowing. Think of the rings as being in the shape of a "C" with the opening of the C toward the back. (A tracheotomy is a procedure to open the trachea between two of these C-rings by inserting a temporary breathing tube.) The trachea then branches into two bronchi.

5. **Bronchi** (two of them leading to the two sides of the lungs) also have rings of cartilage like the trachea. Both bronchi have epithelial tissue with mucus-secreting cells and cilia. Airborne dust, bacteria, pollen, etc., get trapped in the mucus, and then the cilia will beat the waste upward into the pharynx, where it can be swallowed into the esophagus. This helps to clean the respiratory system. The bronchi enter the lungs at the <u>hilum</u> (the term used to describe entry and exit sites on organs—usually for nerves, blood, and lymphatic vessels—but in the case of the lungs, the term is also used for where the bronchi enter).

6. **Lungs** are divided into two portions, with the heart nestled in between. The lungs are protected by the rib cage, and the inter-costal skeletal muscles as well as the diaphragm (another skeletal muscle under the lung) help with breathing. Once in the lungs, the bronchi branch out into smaller tubules called bronchioles; these bronchioles get smaller still and end at the alveoli, the sites of gas exchange.
   - Left lung has two lobes and a notch for the heart.
   - Right lung has three lobes. (So there are five lobes of the lungs altogether.)
   - Lungs are surrounded by serous membranes in the pleural cavity.
   - Lungs are located inside the rib cage for protection.
   - Inter-costal skeletal muscles help with breathing, along with the …
   - Muscle under the lungs called the **diaphragm**, which divides the abdominal and thoracic cavities.
   - When you inhale, the diaphragm contracts, moves down, and expands the thoracic cavity (increasing the space to take in oxygen).

When you exhale, the diaphragm relaxes and moves up, helping to push out your breath. (Abdominal muscles are also helping to push the air out by pushing the diaphragm up.)

Inside lung → bronchioles → respiratory bronchioles → alveoli

7. **Bronchioles** (smaller tubes in lungs). The primary bronchioles lead to the secondary bronchioles, which lead to the respiratory bronchioles (tubules getting smaller along the way). Think of an upside-down tree with the bronchi being two trunks leading to branches leading to twigs with leaves at the end (the leaves are the alveoli). There are no cilia at the ends of the bronchioles; the only defenses left here are the macrophages in the alveoli, which will fight pathogens that have made it past the mucus and cilia along the way. There is cartilage in the primary bronchioles to hold them open, but as the tubules get smaller, the amount of cartilage decreases. The amount of smooth muscle, however, increases in the smaller tubules. If the smooth muscles get irritated, they will contract and it becomes harder to breathe as the tubules are decreasing their diameter or closing altogether (this is what occurs with people with **asthma**). The hormone epinephrine helps to relax the smooth muscles, which allows greater airflow. (We release epinephrine from the adrenal glands into the blood when we exercise or when we have a fight-or-flight response). Asthma inhalers usually have a derivative of this hormone (such as albuterol), which will relax the smooth muscles in the bronchioles to allow a person who's having an asthma attack to breathe easier.

8. **Alveoli** are tiny air sacs that bulge out from the respiratory bronchiole walls. They are clustered in alveoli sacs (think of a bunch of grapes). **This is where gas exchange occurs**. Each alveolus has a single layer of epithelial tissue with a thin respiratory membrane between them and the capillaries that surround them. Diffusion of gases requires a moist environment, so there are special cells, called septal cells, which produce surfactant in the lungs. Surfactant moistens the alveoli, prevents the walls of the alveoli from sticking together, and promotes the uptake of oxygen. When oxygen makes its way down to the alveoli, it takes the place of the hydrogen ions bound to the hemoglobin. The hydrogen ions leave the hemoglobin to recombine with bicarbonate to make carbonic acid, which then breaks apart into carbon dioxide and water. Remember the equation from above?

$$(CO_2 + H_2O \Leftrightarrow H_2CO_3 \Leftrightarrow \mathbf{HCO_3^-} + H^+)$$

The reverse process is occurring before the CO_2 is expelled from the alveoli out the bronchioles $\rightarrow$ bronchi $\rightarrow$ trachea $\rightarrow$ larynx $\rightarrow$ pharynx $\rightarrow$ nose or mouth.

Reminders:

- Cells all over the body make CO_2 during cellular respiration (to produce energy in the form of ATP). The carbon dioxide from the cells diffuses into capillaries and is carried in the blood (attached to hemoglobin or converted to bicarbonate), transporting it to the respiratory system to be expelled (only 7% CO_2 remains in the plasma of the blood).
- Each lung has 150 million alveoli. (If you stretched out all of the alveoli into a single layer, it would be the surface area of a tennis court!)
- Capillaries next to the alveoli move CO_2 out and O_2 in.

**Gases move down their concentration gradient in the alveoli. For example, there is a higher concentration of oxygen inside the alveoli when we breathe in and a lower concentration of oxygen in the deoxygenated blood in the capillaries of the lungs so the oxygen will move from the alveoli to the capillaries.

Ventilation refers to the mechanism to get O_2 in and CO_2 out. Every time we breathe, ~2 cups of air (with 21% oxygen) enters or leaves the body. We usually breathe about 4 liters/minute. Not all of this air makes it to the lungs, however. Each breath delivers about 350 ml of air to the alveoli, and the oxygen from that amount diffuses into the capillaries. The other 150 ml stays in the larynx, trachea, bronchi, and bronchioles, and can exit with the CO_2 during exhalation (this is the reason you can deliver oxygen via CPR to a person suffering from respiratory arrest). Vertebrates use ribs, muscles (diaphragm and inter-costal skeletal muscles), and lungs for ventilation. Different organisms use different methods for ventilation; some amphibians can diffuse O_2 and CO_2 directly through the skin, insects have pores (spiracles) in their sides that the air enters through into tracheal tubes, and fish open their mouths and the oxygen moves from the water into the capillaries in a counter-current exchange. In vertebrates the oxygen that enters the blood is carried by hemoglobin (oxyhemoglobin complex) all over the body for cellular respiration.

The brain monitors ventilation rate:

<u>Medulla oblongata</u> and <u>pons</u> in the brain stem (lower rear tissue of the brain), as well as some chemo receptors in the arteries monitor the acidity (concentration of hydrogen ions) of the blood and control the neurons that stimulate the diaphragm and inter-costal muscles on the ribs to <u>contract</u>. The diaphragm pushes down during inhalation as the thoracic cavity expands.

- In between nerve impulses the muscles relax, the thoracic cavity goes back to the original position, and air is forced out of the body during exhalation.
- If the blood is too acidic (signifying too much CO_2 in the blood), the brain will signal for increased ventilation in order to expel more CO_2 and increase oxygen levels. For example, when we exercise, our cells need a lot of energy for the muscles to contract, so there is a high rate of cellular respiration to make ATP. This means the cells make more CO_2 and therefore the body needs to increase ventilation to both get the CO_2 out as well as to get more oxygen in. We breathe harder and faster when we exercise.

- If our cells run out of oxygen during exercise, we switch to anaerobic respiration, but we make lactic acid in the process. This also raises the acidity of the blood, causing us to breathe more deeply in order to expel CO_2.

Disorders of Respiratory System

Number-One Threat is **TOBACCO** (we already went over this at the beginning of the chapter).

Bronchitis—When air pollutants, bacteria, tobacco, etc., increase mucus production, it can interfere with cilia functions or cause cilia death and inflammation or destruction of epithelial lining in the airways. This causes scar tissue on the insides of the airways, which causes airways to constrict, and there is not as much room for O_2 and CO_2 to move in and out of the airways.
**Smokers have frequent colds, bronchitis

Emphysema—Airways lose structural support and alveoli become weak and can break, which makes gas exchange very difficult. Bronchi are continually clogged by mucus. Emphysema is most commonly caused by smoking, but it can be due to environmental pollutants as well.
- When alveoli start to break down they become surrounded by <u>fibrous tissues</u>; this inhibits gas exchange.
- Genetic factors are sometimes the cause as well. People who lack the gene that codes for a protein to stop a tissue-destroying bacterium get emphysema if they are infected with that bacterium. This kind of emphysema can be treated only with antibiotics.
- The disease usually progresses over 20–30 years, and lungs are permanently damaged.
- People with asthma are twelve times likelier to develop emphysema.

Asthma—A disorder of the bronchioles. The smooth muscles in the bronchioles contract in spasms, making it difficult to breathe. Asthma also causes more mucus secretions, which clog the constricted airways even more. This doesn't permanently damage the lung tissue like emphysema and cystic fibrosis, but asthma can be very hard to live with and can also be fatal. Roughly 10 million Americans have asthma and it kills about 5,000 people a year in the United States!
- Asthma can be triggered by allergens (dust, pollen, dairy, shellfish, pet dander) and viral infections, as well as weather changes and exercise.
- Aerosol inhalers help dilate airways to make it easier to breathe. Asthma medications often have steroids that have some side effects such as stunted growth, hyperactivity, and weakened immune system, so you definitely don't want to take more medication than you need to, but the medication can quickly remedy asthma symptoms and can save lives.

Sinusitis—inflammation and swelling of the sinuses. Sinuses are cavities in the skull lined with mucus membranes. The inflammation is caused by bacterial, viral, or fungal infection and results in swelling and increased mucus buildup, which causes pressure and clogged sinuses. Sinusitis is usually caused by a cold that moves into the sinuses; some cases must be treated with antibiotics or steroids.

Cystic fibrosis—an autosomal recessive genetic disorder resulting from a defective gene that controls consistency of mucus in the lungs. A person with cystic fibrosis produces thick, sticky mucus that traps bacteria and slows airflow. Treatment is physical therapy and drugs to make the mucus more fluid.

Apnea (Abnormal Breathing Patterns)—the mechanism for sensing CO_2 is not working properly and therefore the muscles aren't triggered to help with ventilation. Many people suffer from sleep apnea, which can range from very mild (missing a few breaths periodically; symptoms are often snoring and restless sleep) to severe (some people need to sleep with a breathing machine to ensure getting enough oxygen). The body doesn't store much oxygen, so cells start to die after only three minutes without oxygen (the body will suffer brain damage) and death occurs a few minutes later.

Pneumonia: Caused by bacterial or viral infection.
- Starts in your nose or throat and travels to the lungs.
- Once the infection is in the alveoli, it stimulates them to secrete mucus to try to eradicate the bacterium or virus, but excess fluid in alveoli reduces gas exchange.
- Lung tissue becomes inflamed and fluid builds up, making it very difficult to breathe.
- Can be treated with antibiotics such as penicillin.

Tuberculosis: Caused by bacteria passed in airborne droplets (usually through a cough or sneeze). It destroys patches of lung tissue, and can be treated with antibiotics, but is becoming antibiotic-resistant. The infection can spread to other parts of the body by traveling through the lymphatic system to any organ.

Lung Cancer: Kills more people than any other type of cancer (one-third of all cancer deaths are from lung cancer). Death often occurs within five years of diagnosis so it is a relatively quick-killing cancer. Other cancers can move to the lungs and cause death as well. For example, if cancerous growth in skin tissue gets into the blood vessels, the cancer cells can move to the lung tissue and start growing in there as well. When a cancer moves from one area to another, this is called **metastasis,** which we will discuss further in Chapter 18. Tobacco is the number-one cause of lung cancer.

Chronic obstructive pulmonary disease (COPD) refers to a group of diseases that have to do with damage to the lungs, such as having emphysema, asthma, and chronic bronchitis, all of which obstruct airflow. In the U.S., 120,000 people a year die of this disease. It usually starts with emphysema. Cigarette smoke damages alveoli, forming holes that can't be repaired. The alveoli become filled with fiber, losing their elastic quality, inhibiting gas exchange. Treatment can be by bronchodilators, supplemental oxygen, or lung transplant if very severe.

Choking: Occurs when food gets past the epiglottis and goes down the trachea instead of the esophagus. If you can't get air in or out, you can suffocate within minutes. The Heimlich maneuver can help dislodge the food by elevating the diaphragm muscle in order to force the air up, and hopefully the food will come out with the air.

Drowning: Similar to choking, but this is when water enters the airways and then the lungs. If you can't expel the water quickly, the lungs can't get enough oxygen and death occurs from suffocation (asphyxia) due to not enough oxygen in the brain (cerebral hypoxia) or heart (causes a heart attack). Drowning is the second leading cause of death in children 12 and under (after car accidents) in the United States.

Ear Infection (Otitis Media): When bacteria enter the ear and cause an infection, inflammations will cause the middle ear to fill with fluid, leading to painful stretching of the eardrum. The eardrum can rupture as bacteria within the trapped fluid multiply. Ear infections are usually treated with antibiotics. Bacteria normally enter through the nasopharynx to the Eustachian tube and then finally the middle ear. Children suffer from ear infections more than adults because the tube is horizontal in children, so fluids in the mouth travel to the Eustachian tube easily. As facial bones expand, the tube becomes vertical, and it's harder for the bacteria to make it to the middle ear.

Sore Throat (Pharyngitis): Usually caused by a viral infection that results in inflammation at the back of the throat between the voice box in the larynx and the tonsils. If it is due to a virus, it is often accompanied by other cold-like symptoms and cannot be treated with antibiotics. Be sure to get plenty of rest and vitamin C, drink warm fluids, gargle with salt water, and use pain medication and/or lozenges as needed.

Strep Throat: A sore throat caused by the bacteria Group A Streptococcus. If you have a persistent sore throat often accompanied by white patches at the back of the throat and a fever, be sure to get checked for Strep. Strep throat should be treated with antibiotics since it can lead to rheumatic heart disease or scarlet fever (named for the rash and very high fever that develops), both of which can cause death.

Other Respiratory Associations:

Hiccups—spasms of diaphragm that force air to be pushed out of the lungs, and the epiglottis closes quickly, causing a "hic" sound. It is an involuntary reflex and can be caused by many things. (Fetal hiccups are common too and are associated with fetal development as the phrenic nerve [which triggers the diaphragm] is myelinated prior to 28 weeks. After 28 weeks the hiccups are more related to lung development.)

Hyperventilation—breathing in and out too much/too fast can cause a drop in both CO_2 and O_2 levels, which can result in fainting or loss of consciousness if the oxygen levels become too low in the brain. (Once you lose consciousness, your breathing will go back to normal.)

"Getting the wind knocked out of you"—this is when the diaphragm is temporarily paralyzed, which stops you from breathing. (This happens if a sudden force in the abdomen area blocks the solar plexus nerves so that they can't trigger the diaphragm muscles to contract.)

Sneeze—air forced out of lungs quickly through the nose and mouth due to irritation in the nasal mucosa. This helps to expel foreign particles along with mucus. (Histamines, from the defense system, help trigger this reflex response.)

Cough—another reflex to help clear foreign particles (microbes, dust, pollen, and other irritants) from the respiratory passages. The cough reflex starts with inhalation and then forced exhalation with a surge of air from the lungs against a closed glottis; the glottis then opens and the air is violently expelled.

Laughter—is an expression of happiness regulated by the brain to share this emotion with others. Laughter can be contagious and can help bond social groups together with positive feedback. The sound is due to the epiglottis constricting the larynx. It is a universal language and occurs long before we can speak (babies start to laugh within a few weeks of birth). Laughter and crying can both help deal with stress by counteracting cortisol and adrenaline (our stress hormones). Sometimes we even cry and laugh at the same time, and we're not sure why. We can get sore muscles from laughing, since we contract the diaphragm and abdominal muscles during laughter. The sound, readiness to laugh, and sense of humor are often inherited. The hippocampus and amygdala in the limbic system of the brain are the areas responsible for triggering laughter and other emotions. Laughter is good for your health since it helps relieve stress, increases blood flow, and helps reduce inflammation and platelet accumulation (which reduces the risk of blood clots). Even just smiling can help your well-being, since this triggers the release of serotonin in the brain, which is a "feel-good" neurotransmitter. Try walking around on campus with a big smile and see if it enhances your mood; some people might look at you funny, but I bet you'll also make a bunch of other people smile too.

Yawn—reflex of inhaling air, stretching eardrums, and then exhaling. It is associated with being tired, bored, or stressed, but it is also contagious and can be triggered by watching someone else yawn, thinking about yawning, or reading about yawning. (Are you yawning yet?) There are many theories about yawning, including the idea that yawns help us to stay alert by increasing oxygen levels, help regulate gas levels, regulate brain and body temperature, and increase blood pressure and heart rate, or that the reflex is a leftover adaptation of maintaining social cohesion in a group. None of these theories have been proven, so your guess is as good as mine about why we yawn. I've yawned so many times writing this paragraph that I'm ready to go to sleep!

Chapter 10

Digestive System

I n the previous chapter, we saw how the respiratory system supplies the cells with the oxygen needed for cellular respiration. If your cells don't get enough oxygen, they die. The same is true for nutrients. The digestive system is critical for acquiring a variety of nutrients required for metabolic functions and biosynthesis. (For example, do you remember what the other very important ingredient is to make energy other than oxygen? Glucose!) That is just one important nutrient we need to consume in order to survive. We need all of the organic molecules discussed in Chapter 2 to build and maintain healthy bodies. We take in nutrients in food and beverages and break them down into tiny particles in our digestive tract (also called the gastrointestinal tract). The blood then absorbs the organic molecules, water, vitamins, minerals, electrolytes, etc., from the digestive system, and delivers it to the cells all over the body. Meanwhile the indigestible material (such as cellulose), along with excess fats and cellular debris (dead cells, bacteria, and toxins), are excreted from the body. The digestive tract is a very complex system involving many different organs, glands, and enzymes. (It is also a very long system, measuring 21–30 feet in an adult!) This system is closely related to many other systems as well. The circulatory system, as I just mentioned, picks up the nutrients to take to other locations in the body. It is also delivering oxygen to the digestive system's cells and taking away carbon dioxide. The hepatic system (liver) detoxifies the blood after it is picked up from the digestive system. The nervous system regulates hunger and thirst, as well as all of the smooth muscles that are constantly contracting to push the food through the tract, and the endocrine system (hormones) monitors appetite as well as levels of glucose, glycogen, and fat in the body. Let's start by going over the main functions of the digestive system and then we'll cover the organs, glands, enzymes, and hormones, following nutrients down the tract to see what happens along the way. We will end with disorders of the digestive system. The following chapter will focus on the nutrients themselves, with some issues related to nutrition and health.

Five Main Components of the Digestive System:

1. **Mechanical digestion** is the process of taking in (ingesting), breaking up, and moving food down the digestive tract. It involves the movement of mouth, teeth, tongue, and muscles—making smaller particles with more surface area for the enzymes to work on.
2. **Secretion** of digestive enzymes from digestive organs and accessory glands.
3. **Chemical digestion** of food with the help of enzymes.
4. **Absorption** of nutrients from the digestive system (across the stomach or mostly the small intestines) into the capillaries of the circulatory system to be delivered to other parts of the body.
5. **Defecation** is the elimination of undigested materials and waste such as cellulose, fibers, cholesterol, and excess water (although our systems try to retain most of the water we consume with our foods, reabsorbing it in the colon).

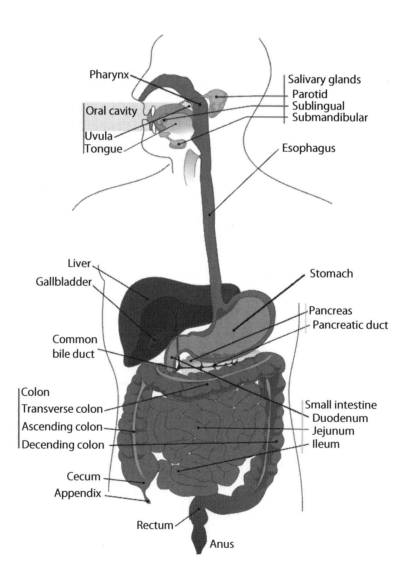

Teeth: Help break up food. Adults have thirty-two teeth, covered in <u>enamel</u>, which is the hardest material in the body. Bacteria in the mouth release acids as they feed and grow (colonies are highest twenty minutes after a meal) and the acid eats into the enamel on the teeth. You can feel the decay once it gets down to the pulp where the nerves are. (That's when you know you need to get a cavity filled!) Bad breath is also from bacteria in mouth.

Tongue: Muscle with receptors (for taste, texture, and temperature) that force food into the pharynx. The tongue also secretes <u>lingual lipase</u>, an enzyme to start the chemical digestion of lipids (breaks down triglycerides).

Saliva: (mostly H_2O, with ions, <u>salivary amylase</u>, mucus, and lysozymes) Secreted from salivary glands under the tongue and near the ears (mumps is a disease that affects these glands). We produce about a half gallon of saliva a day! Saliva starts the digestive process and has many other functions as well:
- Water and ions help to moisten and buffer food.
- Salivary amylase starts to break up some of our ingested carbohydrates. (Starch is broken down into maltose, which will be broken down into glucose by the enzyme maltase in the small intestines. Other carbohydrates, such as sucrose and lactose, will be broken down later in the small intestines as well.)
- Mucus lubricates the chewed food, which is now called a bolus and is ready to be swallowed.
- Lysozyme enzymes digest some bacteria ingested with food.

Pharynx: Otherwise known as the throat, is where the food goes once it is swallowed. It is about five inches long and leads to both the larynx (part of the respiratory system) and esophagus. Remember that this is also the location of the <u>tonsils</u> (to protect again pathogens ingested with food) and the <u>uvula</u> (to stop food/liquids from going into the nose). The bolus doesn't enter the respiratory system, because when we swallow, a flap of cartilage called the **epiglottis** closes off the larynx to divert the food into the esophagus. (We talked about the epiglottis as part of the respiratory system, but it is also directly related to the digestive system.)

Esophagus: Lined with smooth muscle, which rhythmically contracts in a wavelike movement called **peristalsis**. This pushes the food (bolus) down to the stomach. A muscular valve called a **sphincter** regulates the amount of bolus that enters the stomach and then squeezes shut to keep food or fluid from flowing back up into the esophagus. When this esophageal sphincter doesn't close properly, some of the acidic chyme and hydrochloric acid (HCl) can go from the stomach into the esophagus and cause "heartburn" or gastroesophageal reflux disease (GERD).

Liver: An accessory organ in the digestive system, which I will cover in more depth after explaining all of the parts of the actual digestive tract. For now, know that the liver is located above the stomach and the esophagus sits behind the liver to connect to the stomach. One of the main digestive functions of the liver is to make **bile** to help break down fats. The bile is stored in the **gallbladder** and then secreted into the small intestines through the bile duct.

Stomach: J-shaped muscular sac with folds (rugae) that can expand to hold two liters of food, now called chyme.

- The lining of the stomach secretes 2 liters of gastric juice per day! Gastric juice is composed of hydrochloric acid (HCl), which helps with digestion and killing pathogens, pepsin (an enzyme that digests protein), gastric lipase (to break down fats), and mucus. The pH of the stomach is very acidic (pH = 2) due to the gastric juices.
- The alkaline mucus in the stomach helps protect the stomach lining. The lining of the stomach is continually degraded due to the acidity of the gastric juices and needs to be constantly regenerated through mitosis (cell division). We replace the lining of our stomachs roughly every two days! The mucus helps decrease the risk of stomach ulcers.
- The stomach also secretes the hormone **ghrelin**, which is an appetite stimulator. When the stomach is empty, ghrelin triggers your hypothalamus in the brain to signal that you are hungry. (You can hear your stomach muscles churning this organ more clearly when it is empty—that's what the growling-gurgling sounds are!)
- Another hormone secreted from the lining of the stomach is called **gastrin**. This hormone triggers the stomach to produce gastric juices and helps with the lining replacement.
- Chyme is the name for food in the stomach being broken down by gastric juices. After 2–4 hours the stomach has emptied the chyme into the small intestines (small squirts at a time).
- The stomach has three layers of smooth muscle—circular, longitudinal, and oblique. This helps with the mechanical digestion of churning up the chyme. (The rest of the GI tract has just two layers of muscle, no oblique.)
- Pyloric Sphincter: Squeezes one teaspoon of chyme at a time into the small intestine to be processed. Water and alcohol are the first molecules to be absorbed by the small intestines; it happens very fast on an empty stomach, so you feel the effects of alcohol quicker if you haven't eaten. (Alcohol travels to your brain in the blood.) Alcohol can also be absorbed directly from the stomach into the blood, but ~80% of the alcohol in the system will go to the small intestines and then the blood.

Pancreas: (under stomach) A very important accessory gland, which has both exocrine and endocrine secretions. It is called an accessory gland because no digestion actually takes place in the pancreas (the chyme does not go there), but it provides secretions that help with the digestive process.

- The pancreas secretes many enzymes from **exocrine glands** into the duodenum of the small intestine (via the pancreatic duct) to break down different components of the chyme.
- Secretes insulin and glucagon (hormones) **from endocrine glands** into the blood to regulate glucose levels. Insulin is secreted from pancreatic beta cells when blood sugar is high to help muscle and liver cells absorb glucose for cellular respiration and storage. Insulin also acts as an appetite suppressor when blood sugar is high. Glucagon is secreted from pancreatic alpha cells when blood sugar is low since it increases the level of sugar in your blood by converting glycogen to glucose in the liver.
- The pancreas also secretes buffers (bicarbonate) to neutralize the chyme as it enters the small intestines from the acidic stomach.

Small Intestine: Has the very important job of breaking down the macromolecules into smaller building blocks and then absorbing the nutrients into the blood. The small intestines have hundreds of millions of projections called **microvilli** to increase the surface area for absorption. (The small intestine is about 2,800 square feet—the size of a tennis court!) Each villus has circulatory and lymph vessels to move nutrients out of the small intestines. The water-soluble nutrients such as glucose, amino acids, minerals, and vitamins B and C go directly to the circulatory system. The fat-soluble material such as lipids go to the lymph first and then are transported to the blood from the lymph. Once the chyme is in the small intestines, it stimulates the release of the hormones **cholecystokinin (CCK)** and **secretin**. CCK stops the stomach from emptying, and secretin decreases gastric secretions, so both of these hormones ensure that the chyme is held in the stomach long enough to sufficiently break down the particles before entering the small intestines. Secretin stimulates the stomach to produce pepsin, the pancreas to produce digestive juices, and the liver to produce bile so it is working on three different digestive organs. The small intestine will also secrete the hormone **PYY** if there is a lot of chyme in the small intestines. This hormone will act on the hypothalamus of the brain and dampen the desire to eat (acting as an appetite suppressor like leptin). Reminder from the immune system chapter: **Mucus-Associated Lymphatic Tissue (MALT)** is located in the small and large intestines and the appendix. Half of the body's lymphocytes and macrophages are in this tissue to fight off pathogens in the digestive tract. (If there are too many pathogens for the MALT tissue to handle, the body will resort to flushing methods to get them out, such as vomiting and diarrhea.)

Duodenum (10 ft long): The first part of the small intestines receives secretions from the gallbladder and pancreas in order to break the nutrients into smaller particles that can enter the circulatory and lymph systems. It starts breaking down chyme using the enzymes produced in the pancreas and bile produced in the liver (and stored in the gallbladder). The bile duct and pancreatic duct connect together before joining with the duodenum.

Jejunum (3 ft long): The second part of the small intestines is still involved with the digestion of chyme, and now the nutrients are moving from the microvilli into the vessels in a process called **absorption**.

Ileum (6 ft long): More absorption of nutrients into capillaries and lymph vessels occurs here as well as the delivery of unabsorbed material to the large intestines through the **ileocecal sphincter**.

Let's take a closer look at the **layers of the small intestine**, since it is the main absorption area of the digestive system:

The tissue layers in the digestive system work together to create peristaltic waves to send the bolus or chyme along the gastrointestinal tract. The four layers of the GI tract are the same (with the addition of oblique muscles in the stomach), but only the small intestines have microvilli on the mucosa layer. Movement of the tongue when we eat and drink helps to stimulate the swallowing reflex, which then stimulates the muscularis layer to contract in wavelike peristalsis, moving the food down the GI tract.

Looking at the layers from the inside out: the center is called the lumen, where the chyme passes through. The first inside layer is the …
1. Mucosa Layer: Millions of villi and microvilli to increase surface area for absorption (at capillaries)
2. Submucosa: Connective tissue with glands, nerves, blood vessels, and lymph vessels
3. Muscularis: Smooth muscles (circular and longitudinal)

4. <u>Serosa</u>: Outside layer is a very thin serous membrane that keeps tubes in place and protects against friction

The **appendix** is located at the curve where the small intestine funnels into the large intestine. It has no digestive function, but has defensive cells (white blood cells) to combat bacteria consumed in the food. When the appendix is infected (can be caused by stool, bacteria, or a blockage of mucus), it usually needs to be removed before it inflames so much that it bursts, spreading the infection in the abdomen. The surgery to remove the appendix is called an **appendectomy**. Although the tonsils and appendix both play roles in the immune system, we can survive without them.

Large intestine (also called the **colon**) is approximately 5 feet long! It has the major function to reabsorb water, minerals, and vitamins, and then send the undigested material (mostly cellulose and fibers) and excess water and fats out through the anus. Chyme can remain in the colon for 3–10 hours (remaining in the last part for the longest amount of time). The colon also has anaerobic bacteria that live there and help break down undigested material (they make enzymes that help digest complex carbohydrates that we can't digest on our own). The bacteria in the colon are also very important, since they produce vitamin K and folic acid (vitamins we can't make ourselves but that we need), as well as fatty acids. These bacteria also give us gas when they are fermenting solids and they are the cause of the smell of feces. There are four sections of the colon (with sphincter muscles along the way helping to keep food traveling in the right direction).

The <u>ascending</u> portion of the colon is on the right side of the body connected to the ileum by the cecum "blind pouch" (which is also connected to the appendix). The colon moves across the body from the right side to the left along the <u>transverse</u> portion and then heads down with the <u>descending</u> segment. It curves to the middle of the body in the <u>sigmoid curve</u> portion, and then connects to the **rectum**. Sending chyme to the rectum stimulates expulsion of waste out the **anus** (which is the end of the digestive system). Voluntary skeletal muscles control the anal sphincter.

Waste (feces) is composed of water (~75%), undigested materials (~8%), bacteria (another 8%), and the rest is cholesterol, dead digestive cells, salts, and bilirubin from bile (the color is due to the bile/bilirubin). The waste is stored in the rectum until it is ready to be expelled.

Insoluble fiber is very important for our diet because it helps to expel waste. This is important not only for preventing constipation but also because old waste can break down into cancer-causing chemicals and can result in colorectal cancer, which is one of the deadliest cancers. Guidelines for "regular" bowel movements can range from 3 per day to 3 per week, but there are also healthy individuals who fall outside these ranges on either end. The key to having regular bowel movements is to have a healthy diet (plenty of fruits, veggies, and water) that helps to maintain the proper elimination of waste without constipation or diarrhea.

Okay, now we're going to return to the liver in order to cover some of the main functions of this amazing organ. It is a critical accessory organ to the digestive system since the digested material that is absorbed from the small intestines travels in the blood to the liver for further processing.

Liver: (above the stomach) The largest organ in our body other than the skin. It weighs about three pounds, and makes hundreds of types enzymes for all of the chemical reactions that take place in liver cells all the time. It is critical for the digestion of fats, but has many other functions, such as metabolism, storage, and detoxification. Nutrients and toxins come into the liver from the small intestines via the hepatic portal vein. (Remember that this is one of the only veins that does not go directly into the heart—once the liver has detoxified and stored nutrients from the blood, the rest goes out the hepatic vein to the heart.) Out of the hundreds of jobs the liver performs, I will just list the main ones that I want you to know. (As I've already mentioned, the liver is like the UPS of the organ system—responsible for manufacturing, processing, shipping, and storage.)

Main Functions of Liver:

1. **Produces and secretes bile** (~1 liter/day) that emulsifies fats. It is a biological detergent that breaks large fat globules into smaller ones that can then be broken down by pancreatic lipase. Bile also helps rid the liver of excess cholesterol.
   - Hepatocyte cells in the liver make bile, which is composed of salts, water, fats (cholesterol), and other chemicals like bilirubin (a pigment byproduct of the breakdown of red blood cells). Bile is secreted into the gallbladder along with an alkaline bicarbonate solution and excess cholesterol from the liver (which dissolves in the bile solution).
   - Bile is stored in the gallbladder (under the liver) and then secreted into the duodenum of the small intestine when food (chyme) enters the system.
   - The bile duct from the gallbladder connects with the pancreas duct before reaching the duodenum.
   - Ninety-five percent of the bile salts are absorbed in the blood from the ileum and brought directly back to the liver to be reused.
   - Without bile salts, most of the lipids ingested would be excreted as indigestible wastes (and we need these lipids to make triglycerides, phospholipids, etc.).
   - Bile is not an enzyme. It helps with mechanical digestion rather than chemical digestion.
2. **Controls amino acid and ammonia levels** in blood. The liver makes many proteins and ships them to cells to use or uses them for reactions within the liver. If amino acids are not used to make new proteins but instead are broken down into ammonia (which is toxic), the liver cells convert the ammonia into **urea**, which is then sent to the kidneys through the blood, and the urea gets excreted as part of the urine.
3. **Controls glucose levels** and can **convert glycogen to glucose** and vice versa. (When glucose levels exceed 0.1% in blood, hepatocytes remove and store the excess as glycogen.) The hormone glucagon, secreted by the pancreas, stimulates the liver to convert glycogen back to glucose. When the glycogen levels get too low, the glucagon can stimulate the liver cells to synthesize glucose by gluconeogenesis. (Cells use glucose for cellular respiration to make ATP; if they don't need ATP, then the glucose is converted to glycogen and is stored in the liver for later use. When glucose is needed again, the liver converts glycogen back to glucose.)
4. **Removes hormones** that aren't needed any more, which are then sent to the urinary system.
5. **Breaks down worn-out or dead red blood cells**, which get sent to the spleen for disposal.
6. **Stores some vitamins** such as A, D, and E, and **stores iron.**
7. **Stores fat**, or ships it out to the adipose tissue. (Also rearranges the fat molecules to make lipoproteins.)

8. **Fat-storing cells in liver secrete the hormone** <u>leptin</u>—as fat storage increases, leptin suppresses the appetite.
9. **Cholesterol, plasma proteins, and blood lipids** are **manufactured** in the hepatocytes in the liver.
10. **Lactic acid made by muscles is converted to glucose**.
11. **Detoxifies blood** by converting toxins from alcohol, caffeine, cold medicines, etc., into less-harmful chemicals.

Liver cells specialize in helping rid the body of toxins such as alcohol (ethanol) by making enzymes in peroxisomes, which, for example, convert toxic ethanol to less-harmful acetaldehyde, and then acetate, and finally carbon dioxide and water. It takes these enzymes about one hour to detoxify 10 grams of alcohol (~1 alcoholic beverage). Overconsumption of alcohol damages the liver by eventually killing the liver cells, and can lead to different diseases of the liver, such as fatty liver, alcoholic hepatitis, and alcoholic cirrhosis. (Detoxifying alcohol takes up O_2 in the liver cells. Oxygen, however, is also needed for the cells' basic metabolic functions, and if there is a shortage of oxygen for these functions, the cells die.) Drinking can impair brain function as well, not only when you are "drunk," but it can also lead to brain damage in heavy drinkers over time. (We will talk more about alcohol when we get to the nervous system.) Binge drinking is especially dangerous, because if you drink too much too fast, it can lead to heart failure. (Acetaldehyde circulating in blood attracts monocyte immune cells, which can lead to blockages in blood vessels, and heart attack.) On the other hand, some alcohol in moderation (like red wine) can be beneficial to the heart and provide antioxidants to fight against cancer.

Central nervous system control of digestion:

1. Involves the medulla oblongata **and the hypothalamus**. Thoughts and smells of food stimulate the smooth muscles of the stomach to start churning and the stomach lining secretes gastric juice. It also stimulates the secretion of saliva.
2. As the stomach stretches with food, it triggers the "full feeling" from your brain so that you stop eating, triggers muscle contractions for peristalsis to keep food moving through your digestive tract, and triggers the secretion of enzymes from different glands.

There are also **endocrine cells** in your GI tract to help with digestion. For example, when the stomach has protein in it, the hormone **gastrin** is secreted into the bloodstream, which stimulates the release of **pepsin** and **HCl**, key ingredients of gastric juices to help break down proteins. Once the stomach is empty, the hormone **somatostatin** stops HCl from being released. The stomach also secretes **ghrelin**, which is an appetite stimulator. It travels from the stomach to the small intestines and is absorbed into the blood, where it will travel up to the brain. The hypothalamus is the central part of brain that controls hunger. The hypothalamus also receives messages from the hormones **PYY** (released from the small intestines) and **leptin** (made in the fat cells of the liver), to help regulate hunger.

Digestive System Diseases and Disorders:

Colon Cancer: The third leading cause of cancer deaths in the United States for both men and women (lung is number one, followed by prostate for men, breast for women). Colon cancer can be detected by examining the stool for precancerous cells or by a **colonoscopy**, which examines the colon wall for polyps (growths of cancerous cells). We will discuss colon cancer further in Chapter 18.

Malabsorption disorders: Anything that interferes with the small intestine's ability to absorb food.

Lactose Intolerance: When you lack the enzyme <u>lactase</u>, which breaks down lactose (type of sugar from dairy) in the small intestines. Lactase helps the body absorb lactose; people who are lactose intolerant usually feel sick after eating anything with lactose in it.

Cystic Fibrosis: The body doesn't make normal pancreatic enzymes to break down and absorb fats and other nutrients.

Crohn's disease: An autoimmune disease in which the immune system attacks part of the digestive tract, anywhere from the mouth to the anus, but usually in the small intestines. Nutrients are not properly absorbed from the digestive system to the circulatory system, and it can cause weight loss, vomiting, inflammation, pain, diarrhea, and other symptoms. It is often genetic but can be environmentally induced as well. It usually doesn't appear until the teens or 20s but can occur at any time. Men and women are equally likely to suffer from this disease, and smokers are three times more likely to be affected than non-smokers.

Infections and Diseases of GI Tract:

Constipation: Too much fluid is reabsorbed by the colon, and feces are hard to get out. Constipation can cause enlarged rectal blood vessels also known as hemorrhoids. Adding more fiber and water to the diet can often help with constipation.

Diarrhea: When the small intestines secrete more water than the large intestines can absorb and the result is watery stool. It causes dehydration and electrolyte imbalance and is the second-most common cause of death in infants in the world! Bacterial, viral, or protist infections can trigger this condition in the body's attempt to flush the pathogen out. It can also be from malabsorption disorders, stress, and other factors that speed up peristalsis. It is very important to replace fluids and electrolytes after diarrhea.

Giardia: Caused by a protista that gets into the digestive tract through contaminated food or water. This usually leads to diarrhea in the body's attempt to expel the pathogen.

E coli: Bacterium that lives in intestines of cattle, pigs, and humans. When we consume a virulent strain, it can lead to diarrhea, dehydration, and anemia (some strains kill red blood cells). (Many strains are harmless and help us!)

Epstein-Barr Virus (EBV): Causes mononucleosis (kissing disease). It is transferred orally through saliva (sharing food, cups, and utensils, and kissing) or by coughing/sneezing. The virus moves from the saliva through the mucus of the upper respiratory system into the bloodstream, where it attacks B-cells. Leukocyte cells become activated (specifically mononuclear leukocytes, hence the name). This occurs mostly in females ages 10–35. It should be treated with rest and fluids—usually lasts for 2–3 weeks, but fatigue can last a lot longer. EBV is often misdiagnosed as strep throat initially (until the strep test is done) because they both have the similar symptom of white bumps at the back of the throat.

Salmonella: Bacterium in meat/poultry/eggs (food poisoning) can cause septicemia if it enters the blood and spreads bacteria throughout the body.

Gingivitis: Bacteria in gums due to poor oral hygiene can cause gum disease, which is actually a major killer! It not only leads to cavities but is also related to heart disease, stroke, lung disease, diabetes, and premature births. (Eighteen percent of premature births are due to gum disease—hormones during pregnancy can make the mom react differently to bacteria, increasing the occurrence of gum disease and cavities.)

Peptic Ulcers: Ulcers refer to holes in the stomach or intestinal lining, usually caused by Helicobacter pylori bacteria, or excessive alcohol, ibuprofen, or aspirin use. When mucus lining that normally protects the stomach or small intestines is compromised, the acidity in the lumen starts to burn the stomach lining, and pepsin digests proteins in the stomach cells without mucosal protection. It used to be thought that stress caused excess stomach acids, which lead to ulcers, but now it's thought that it has more to do with bacteria in the stomach, or pain medication. If it's bacterial, antibiotics are prescribed; if it's not bacterial, the ulcers will heal over time once the cause is removed.

Heartburn: Also known as gastroesophageal reflux disease, is when the esophageal sphincter does not close properly, and the acidic chyme from the stomach goes back into the esophagus. Medicine can be taken to reduce the acid, and less-acidic food should be eaten. Don't lie down soon after you've eaten if you have heartburn and definitely don't smoke since tobacco exacerbates this condition. Some women get heartburn during pregnancy since the expanded uterus presses against the digestive organs, causing some chyme to enter the esophagus, but this condition is usually temporary and will subside once the baby is born.

Gallstones: There are two types of gallstones; some are formed from excess cholesterol that is secreted with the bile and can form into hard cholesterol balls in the gallbladder; the other type is from stones made from too much bilirubin in the bile, and these are called pigment stones (less common). If the gallbladder needs to be surgically removed due to gallstones, the liver will still produce and secrete bile; the bile duct can enlarge to become the storage area.

Vomiting (Emesis): The forceful expulsion of chyme from the stomach out of the mouth (due to con-traction of abdominal muscles). If the bolus is still in the esophagus, it is called regurgitation instead of vomiting. Both can cause dehydration and electrolyte imbalances. Vomiting can be triggered by the nervous system in cases such as motion sickness, stress, migraines, pregnancy, or psychiatric disorders. It can also be manually forced in cases such as bulimia. Often it is because of pathogens in the GI tract (such as the norovirus that causes the stomach flu). After an upset stomach it is sometimes hard to go back to eating your regular diet. To help settle the stomach, the BRAT diet is recommended: Bananas, Rice, Applesauce, Toast/Tea, and plenty of fluids. (Ginger helps for some people as well—ginger snaps and ginger ale can also be good post-nausea remedies.)

Chapter 11

Nutrition

This chapter is related to the digestive chapter in the sense that we are still discussing the foods we eat, but instead of the physical process of breaking down the nutrients, let's look at why the nutrients are so important. We will first go over what the main types of nutrients are, where to get them, and how they help our metabolic functions. We will then discuss some nutritional recommendations, and how to maintain a healthy body by eating nutritional foods, watching how much we consume, and exercising regularly. **Metabolism** refers to the chemical reactions in the body that break down nutrients and build compounds, such as organic molecules. Metabolism contributes to our growth, survival, and reproduction. We all have different rates for these metabolic reactions, just as we all have different rates for digestion. I will provide some numbers to live by—such as healthy levels for cholesterol, sugar, protein, and fats, as well as recommendations for weight and activity levels in order to maintain a healthy body. The chapter will conclude with some dietary problems and eating disorders.

A **Nutrient** is any compound required by our body that we cannot make on our own and must ingest or have made for us (such as the vitamins that bacteria make for us in the digestive tract). There are two types of nutrients:

Macronutrients (organic molecules) provide the building blocks for ATP, proteins, cellular components, glucose, lipids, whole new cells, etc., and **micronutrients** (vitamins and minerals) often help run reactions and are critical for many processes such as muscle and nerve contractions, metabolic processes, immune system, mental functions, and building tissues.

Three Main Classes of Macronutrients: Carbohydrates, Protein, and Lipids.

1. Carbohydrates are our most efficient source of energy. They can also boost our mood since some carbohydrates can trigger the release of the feel-good neurotransmitter serotonin in the brain. There are two main types of carbohydrates: non-processed carbohydrates and refined carbohydrates. Let's look at these separately here.

Non-processed carbohydrates refer to whole grains, fruits, and vegetables. A healthy diet should contain 75% of these carbohydrates in order to meet our energy requirements since they are loaded with starch and fiber.

Starch is a very important complex carbohydrate that we need to make energy. Starch is found in fruits, vegetables, grains (breads, pasta, rice, cereals), and legumes (peas and beans). We use enzymes to break up starch into the simple sugar **glucose**, which is needed for cellular respiration to make ATP, and to synthesize amino acids, glycogen, and triglycerides.

Plants also contain **cellulose**, the sugar in their cell walls. This is also known as **fiber**. There are two types of fibers: soluble and insoluble. **Soluble fiber** dissolves in water and we can digest it ourselves. It slows the digestive process down, which helps you to feel satiated so that you don't overeat. It also helps with the absorption process in the small intestines, and to lower fats, LDL cholesterol, and insulin levels in the blood. It reduces the risks of heart disease and type II diabetes. **Insoluble fiber** is the kind that we cannot digest. It is equally important in our diet, however, because it absorbs water in the small intestines, which makes waste softer and easier to expel. This helps to clean out the colon and reduces the risk of colon cancer. Bacteria in our large intestine are helping to break down these complex carbohydrates that we can't digest as well. This leads to intestinal gas (so farts are generally a sign of good nutrition). In order to get both types of fiber, it's important to include a variety of fruits, vegetables, beans, brown grains, cereal with bran, oatmeal, nuts, seeds, and whole wheat foods in your diet. (We need **lots** of fruits and veggies in our diet for this reason, as well as for the starch, vitamins, minerals, and sugars they provide).

The other main type of carbohydrates are called **refined carbohydrates**, which are made from refined flour (grains that have been stripped of their nutritional seed coat and are lacking the bran and fiber). Examples include white bread, crackers, cookies, and pastries. These tend to be high glycemic index foods.

Glycemic Index (GI): Measures the refined-sugar levels in blood. High glycemic index foods are digested quickly due to the refined simple sugars and cause spikes in blood sugar levels. When this happens insulin levels go up (secreted from the pancreas), which helps cells absorb sugar quickly and stops the body from using stored fat. The liver converts excess glucose to glycogen (stored in the liver), and then converts nutrients digested in the small intestine into fat in the form of triglycerides. The liver sends the fat out (in vesicles) into the bloodstream and it gets deposited into adipose tissue, which is basically fat cells held together with other connective tissues. When the blood sugar level falls, the insulin levels go down as well, and you feel hungry again.

Since these types of foods are not adding nutritional value to our diet, they are called "empty calories." They can provide a quick burst of energy but they mostly promote over-eating, weight gain, and insulin resistance. Eating too many refined carbohydrates can increase the risk of obesity, type II diabetes, cancer, heart disease, and other diet-related health problems.

There is a variety of **simple sugars**, such as sucrose, fructose, lactose, maltose, and glucose. Our body responds to these sugars in different ways. We're consuming fructose and sucrose in fruits and vegetables but these are healthy to eat since they are also high in fiber, starch, vitamins, minerals, antioxidants, and lots of water. The types of sugars that we should limit are those found in sweets, processed foods, fast foods, sodas, and juices, many of which are sweetened with **high fructose corn syrup** (HFCS).

As mentioned in Chapter 2, high fructose corn syrup usage has increased dramatically in the past 30 years. It is not surprising that the obesity rate has skyrocketed in the United States during the same timeframe. We are not designed to break down artificial sweeteners the same way that we metabolize glucose. (HFCS is from genetically engineered corn.) Rather than using the sugars for energy production, they mostly get converted into fat. HFCS has also been linked to harming the digestive tract, accelerating the aging process, and causing infertility in some people. Try to limit these types of sugars to 32 grams per day. There are 4 grams of sugar in a teaspoon so this means <u>8 teaspoons of sugar a day</u>. (The national average consumed right now is more like 30 teaspoons a day!!!) To give you an idea of how much we are consuming … one can of soda is 9–13 teaspoons of sugar and roughly 200 calories. A specialty coffee drink can be close to 400 calories and can have twice the amount of sugar as a soda! Juices, although they sound healthy because they contain fruits, can also have more calories and sugars than sodas (although I would still pick the juice because then at least you're getting some vitamins and minerals too.) The best choices for beverages are water (of course!), milk (which contains only about 2–4 tsps of sugar and has plenty of protein, calcium, and vitamin D), and lightly or unsweetened tea or coffee.

Americans consume 70–130 pounds of sugar per person a year! That sounds obscene, but it is also one of the main reasons 68% of American adults are overweight or obese. The main causes of obesity and Metabolic Syndrome are eating too much and too often, eating unhealthy foods loaded with fats and sugars, and not getting enough exercise. Both can lead to high blood pressure, high triglyceride and blood sugar levels, lower HDL levels, too much fat around the waist, and insulin resistance, which, as already mentioned above, can lead to increased risk of heart disease, type II diabetes, cancer, stroke, and death.

2. Proteins are incredibly important body-building nutrients, especially for bone and muscle. Proteins found in lean meats are also important for boosting dopamine levels in the brain, which can make you feel more alert and energized. Proteins in food are broken down into amino acids. There are twenty amino acids, but we can manufacture only ten of them ourselves. The other ten amino acids are called "essential amino acids" because we need to get them from our diet. (Some sources list only eight or nine essential amino acids. This is because one amino acid is not needed in adults. There is also some discrepancy due to the fact that one amino acid that the body **can** manufacture depends upon having another amino acid from the diet. For the purposes of this book, we will just call it ten essential amino acids.)

Proteins are not stored for later use like carbohydrates and fats. Instead they are broken down and the amino acids are used to make fuel (energy) or to make other proteins that are necessary for our metabolic functions, immune system, hormones, etc. We especially need proteins when there is tissue damage or during growth. (Remember all of the functions of proteins? Refer to Chapter 2.) When cells die, the proteins within the cells are also broken up and the amino acids are recycled to make other proteins. Once the body has synthesized enough proteins, the excess amino acids will be formed into ammonia and then converted to urea (in the liver) to be excreted by the urinary system. (Excess amino acids are not stored for later use.)

The recommended caloric intake from proteins ranges from 10% to 35%. This number varies greatly depending on a person's developmental stage, health, weight, and activity levels. For example, a child, a pregnant or lactating woman, or someone recovering from surgery or an illness needs to consume more protein than a sedentary adult. On average, a healthy adult should consume roughly 0.8 grams of protein per kilogram of body weight per day. (One kilogram equals 2.2 pounds.) So a 130-lb person would need roughly 47 grams of protein a day versus a 180-lb adult who would need closer to 65 g/day. I know body builders generally consume more protein to add to their muscle mass but you don't want to overdo your protein consumption: no more than 35% of your diet. Consuming too much protein can harm your body by causing dehydration, which can lead to less muscle mass! It can also put excess stress on the kidneys and heart and can lead to kidney stones, heart disease, and colon and liver cancers. To get an idea of how many grams of protein are in various foods we eat: 1 cup cottage cheese = 28 g, 1 cup milk (1%) = 8.2 g, 1 egg is 6 grams of protein, 3 ounces turkey, chicken, salmon, and tuna are all around 20–25 g, and a cup of soybeans (cooked) is 28 g! To see a list of protein content in a wide range of foods, go to <www.health.harvard.edu/usda-protein>.

As babies, we get all of the protein we need from milk. Once we start eating solid foods, there are a variety of sources from which we can get the essential amino acids we can't make ourselves. It is crucial to get all ten essential amino acids, since deficiencies in any one of them can lead to health problems. Remember that the twenty amino acids are being put together in different combinations to make up all of the different proteins in our body, much like the letters of the alphabet are used to make different words. If you are missing an amino acid, then many different proteins will not be made, which can affect numerous bodily functions.

Essential Amino Acids: We need to get these amino acids from our food since we can't make them ourselves.

There are proteins in vegetables, fruits, nuts, seeds, grains, legumes, yeast, freshwater algae, etc., and many of these vegetarian sources will provide you with most of the essential amino acids, but you need to eat a combination of them to get all ten. (Called **incomplete foods**, since they don't have all ten themselves.) Soy and quinoa are two vegetarian sources that have a complete set of the essential amino acids. (Soy comes in many forms, such as milk, tofu, ice cream, burgers, hot dogs, etc., so vegetarians often include some sort of soy product in their diet.) There are many food combinations like corn and beans, red beans and rice, and peanut butter and bread, which will also provide all ten. If you are a vegetarian, do a little research to ensure that you are getting all of your essential amino acids.

Animal foods—meat, eggs, cheese, and other dairy products can provide all ten essential amino acids, so they are called "**complete**" **foods**. Remember that overindulging in animal foods, however, can lead to heart disease, kidney stones, and colon and liver cancers, so it is important to not overeat these types of foods.

3. Lipids are the third essential macronutrients that we need to get in our diets. Phospholipids make up the cell membranes; triglycerides are important for energy storage, cushioning, and insulation; steroids such as cholesterol and hormones are necessary for building cells, communicating between cells, and are used in digestion (bile), etc.; and waxes are for protection. We get some of our lipids from our diet and the rest of them are made by the liver from carbohydrates and proteins.

Fats

- If you consume more calories than you need for energy (usually from carbohydrates, fats, sweets, and alcohol), your triglyceride level will increase and the liver will store the fat in adipose tissue. As I mentioned already, a certain amount of fat is important for cushioning, insulation, protection, thermoregulation, and fuel reserves, but too much can also lead to heart disease. You should limit consumption—especially limit the "bad fats" in your diet. (Triglyceride levels should be under 150 mg/dL = milligrams per deciliter.)
- **"Saturated fats"** have two hydrogen atoms attached to every carbon and are hard to break down. They are known as the "bad fats" and are solid at room temperature. They bind with cholesterol in vessels to add to plaques and increase the risk of heart disease. Examples include butter, meat fat, chicken skin, dairy fats (cheeses), coconut, and margarine. The worst type of saturated fat is **trans fat**, also known as "partially hydrogenated vegetable oil." This fat comes from adding hydrogen to vegetable oil in order to make products last longer (ex: added to a lot of processed baked goods like cookies, crackers, cakes, etc.); these type of fats raise LDL (bad) cholesterol and lower HDL (good) cholesterol levels. Fried foods like donuts and french fries sometimes have trans fats as well.
- **"Unsaturated fats"** refer to fats that have double bonds in the hydrocarbon chains, which make them kink and the fat cannot pack as tightly together—if they have one double bond they are called monounsaturated and if they have more than one double bond they are polyunsaturated. They are all liquids at room temperature. There are many types of unsaturated fats but the ones you need to consume are called **"Essential fatty acids"** (**EFA**s) because we cannot synthesize these lipids ourselves and must get them from our diets. These include the polyunsaturated omega-3 and omega-6 (linoleic) oils. Good sources for these are flaxseeds and fish, and a variety of oils including linseed, canola, sesame, sunflower, corn, and soybean. Omega-3 fats are particularly beneficial for lowering the risk of heart disease, stroke, inflammation, and cancer. They have also been linked to boosting brain function and lowering the symptoms of ADHD and depression. Other examples of foods with good unsaturated fats are olives, olive oil, peanuts, nuts, avocados, seeds, and eggs. Be sure to incorporate some of these "good" fats into your diet (in moderate amounts).

Cholesterol:

- We make all of the cholesterol that we need for hormones, vitamin D, bile, and membranes, but we also consume cholesterol in our diet (from animal fats). Excess cholesterol that can't be processed by the liver is sent into the bloodstream with proteins around it. Too much of this excess cholesterol circulating in our blood can cause health problems depending on our levels of good vs. bad cholesterol.
- Fats (such as triglycerides) and cholesterol are both traveling around in the blood but they are insoluble (can't dissolve) and therefore combine with proteins for easier transport in the body. Different types of lipids + proteins (around the outside) are called lipoproteins. When we talk about our cholesterol levels, we are usually referring to these packages of lipoproteins, of which we have two types: Low-Density Lipoprotein (LDL), and High-Density Lipoprotein (HDL).

LDL (Low-Density Lipoproteins)
- Have more cholesterol than protein: 50% cholesterol, 5% lipids (triglycerides), 25% proteins, 20% phospholipids (good for cell membranes).
- Travels in blood and can clog arteries (atherosclerosis = fatty deposits in walls of blood vessels, which can lead to heart attack and stroke).
- You do not want to have high LDL. Hypercholesterolemia refers to high cholesterol in the blood (high LDL levels). LDL levels should be 100 to 129 mg/dL. (Even lower if you have risk of heart disease: heart disease in family, previous heart attack, smoker, high blood pressure, or diabetes.)
- People with high cholesterol can modify their diets, exercise, or take medicine (statins) to lower cholesterol levels. There are some new drugs prescribed as preventative medicine to reduce the chances of heart attack for people who don't have hypercholesterolemia; however, these drugs increase the chances of diabetes. Always do your research before taking any medications, especially new "preventative medicines."

HDL (High-Density Lipoproteins)
- Have more protein than cholesterol: 15% cholesterol, 10% triglycerides, 45% proteins, and 30% phospholipids.
- Important since HDL is needed to carry cholesterol away from cells. It also helps to carry LDL cholesterol and triglycerides away from vessel walls and deliver them to the liver for elimination and excretion. The liver also converts cholesterol into bile salts to help with digestion of lipids.
- HDL can also be brought to adrenal glands, ovaries, and testes to make hormones.
- Denser than LDL because protein is denser than lipid.
- Good cholesterol goes to liver to be broken down.
- HDL levels are best above 60 mg/dL in order for HDL to help remove cholesterol from blood.
- Healthy diet, aerobic exercise, no smoking, mild alcohol, and adding omega-3 fats and fiber to your diet can help boost HDL levels.
- Total LDL + HDL levels are best under 239 mg/dL. (Per Kaiser Permanente, 2012)

Dietary recommendations (Constantly changing, but these are the current guidelines):

Water: Approximately 8 cups/day (A good rule of thumb is to drink half your body weight in ounces. For example, if you weigh 140 pounds, drink 70 ounces of water. One way to know if you are drinking enough water is to look at the color of your urine. If it is pale yellow you are doing great—if it is concentrated yellow be sure to drink more water.) Water is okay to drink with food despite the myth that it interferes with digestion. Water is the most important liquid in our body—necessary for cell hydration, chemical reactions, regulating temperature, brain activity, etc. We can live weeks without food but can only survive about 3 days without water!

Dark green veggies: 2.5 cups/day. ex: spinach, chard, broccoli…

Orange veggies: 3–6 cups/week: carrots, squash, sweet potatoes…

Fruits: 2 cups/day: Citrus, melons, berries, apples…

Milk products: 2–3 cups/day: milk, cheeses, yogurt …

Grains: 6 ounces/day: Bread, pasta, rice, cereals… (our main energy source), at least 3 of these grams should be whole grains (whole wheat, oatmeal, brown rice, quinoa, barley, whole wheat cereals, popcorn, etc.) 1 ounce = 1 slice bread, 1 cup cereal, or 1/2 cup pasta, rice, or crackers.

<u>Fish, poultry, eggs, beans, nuts, meats</u>: 5 ounces/day.

<u>Oils</u>: 24 grams/day: olive oil (good antioxidant), canola oils …

(You can check the government agriculture website <<u>www.choosemyplate.gov</u>> to see how many calories a day you need. This outline is based on about 2,000 calories a day.)

Basically, if you have a plate of food and 50% is from fruits and veggies, 25% is from grains (whole is the best), and the remaining 25% is lean proteins, you're doing great (plus your milk and water). When I calculated a sample recommended diet of 300g carbs, 65g fats, and 50g proteins, the breakdown was similar: 72% good carbohydrates, 16% fats, and 12% proteins.

The next type of nutrients that we must consume (because we can't synthesize them ourselves) are the **micronutrients**, **vitamins**, and **minerals**. They are just as critical to our function and survival as macronutrients. We need to make sure that we have a diverse balanced diet (with plenty of fruits and vegetables) so that we are getting all of the micronutrients necessary for life.

Vitamins: <u>organic</u> substances (they contain carbon) that are required for growth and survival. We generally can't make them, so we must consume them in our diet. (The only exceptions are vitamin D, which we can make with the help of the Sun, and folic acid and vitamin K, which we can get from bacteria in our digestive system.) There are more than 20 different vitamins, but some of the main ones are A, B, C, D, E, and folic acid. All are important for different functions. I'm just going to cover a few critical vitamins and minerals in this text.

Vitamin B$_{12}$: Involved in the metabolism of every body cell. Vitamin B is good for DNA synthesis and regulation, fatty acid and energy production, and manufacturing neurotransmitters. It is also especially important for the nervous system and the formation of blood in red bone marrow. Found in meat, cheese, eggs, and liver. It is often taken for insomnia, depression, metabolic function, and to help boost alertness and focus.

Vitamin C: From fruits and veggies, needed for bones, teeth, collagen, cartilage, and immune system. If you don't have enough, you can get scurvy, which can cause death.

Folic Acid (aka Folate and B$_9$): Found in dark green veggies, grains, and yeast. This vitamin is important because it is a coenzyme to help enzymes build molecules (especially nucleic acids for DNA/RNA). Bacteria make folic acid for us in the digestive tract, but it's good to get extra in your diet as well. Pregnant women should make sure to get enough folate since it is also very important for fetal development. Folate deficiencies can cause developmental problems for the fetus, and in adults it can cause cognitive problems that have similar symptoms to Alzheimer's.

Vitamin K: One type is produced by bacteria in our large intestine and we can get the other type from dark green leafy vegetables. Vitamin Ks are necessary for proper blood clotting, strengthening bones, and keeping the heart healthy.

Vitamin D: A lipid vitamin synthesized in our skin! The Sun activates the conversion of cholesterol in skin to vitamin D, which is then transported to the small intestines where it promotes calcium absorption. Rickets is the softening of bones due to vitamin D deficiency. We can also consume vitamin D in our diets—milk and fish are great sources of this important vitamin. (Vitamin D is also thought to help fight cancer.)

Many vitamins are antioxidants that help to neutralize free radicals, important for reducing the risks of cancer. I will discuss this further in Chapter 18, but for now I wanted to mention that vitamins A, C, D, E, and K are all effective antioxidants, along with **phytochemicals** like flavonoids found in cocoa, tea, wine, and berries.

Minerals: As opposed to all of the macronutrients and vitamins, minerals are <u>inorganic</u> substances (they do <u>not</u> have carbon). However, they are just as essential for the body's metabolic functions. Some of the main minerals we need are iron, potassium, calcium, sodium, magnesium, iodine, and zinc. Here are some details about a few of these.

Iron—Helps hemoglobin formation (hemoglobin is the protein inside red blood cells that carries oxygen in the blood). The composition of this protein has the mineral iron right in the center. If you don't have enough iron in your system, you can't make enough hemoglobin and therefore your blood will not carry around enough oxygen. This causes anemia. Iron is found in red meats, poultry, fish, lentils, beans, dark leafy greens, and fortified breads and breakfast cereals.

Zinc—One of the most important minerals since it is necessary for the proper functioning of roughly 100 enzymes assisting many different metabolic reactions. It is especially important for cell division, cell growth, and immunity. It is found in many foods such as red meats, oysters, lobster, beans, nuts, seeds, and whole grains.

Calcium—A critical mineral for the proper functioning of muscles and nerves. It also hardens bones and teeth and has a role in cardiac contraction. If you don't have enough, you can develop nervous system, skeletal, and muscular disorders. Good sources for calcium are dairy foods, oranges, spinach, seaweed, almonds, broccoli, and beans. Vitamin D is necessary for proper calcium absorption in the small intestines.

<u>Nutrient Recommendations:</u>

<u>Reduce bad saturated fats</u> (animal and dairy fats) since they increase the risk of heart disease and cancer. Make sure to get enough of the <u>good unsaturated fats</u> (oils, nuts, olives, avocados, seeds). <u>Increase omega-3 fats</u> (flaxseeds, green veggies, fish, and walnuts) to reduce the risk of cancer. **Limit total fats to 65 grams per day** (saturated fats should be less than 20 g/day). Keep **carbohydrates to around 300 grams per day** and **dietary fiber around 25–28 grams per day**. (Grams of protein can really vary depending on your activity levels, developmental stage, body mass, etc., but the recommended caloric intake from **proteins ranges from 10% to 35%**. (Fifty to 75 grams per day is the general range for a 140-pound person.)

Get plenty of **vitamins** and **minerals** and foods with **antioxidative** properties to help neutralize **free radicals**, boost the immune system, and reduce the risk of cancer. Multivitamins have vitamins E, C, and A, all of which help remove free radicals. Everyone's body generates millions of free radicals every minute. Free radicals are unstable oxygen molecules that are made all the time in our body during cellular respiration. We break off the pieces of oxygen that we need to make ATP and leave behind the free radicals. Free radicals are also inhaled in cigarette smoke and in pollution. They damage the skin, and they cause internal inflammation, which can lead to heart, lung, and GI diseases, and well as some cancers. Foods rich in antioxidants include blueberries, cranberries, red wine*, dark chocolate, dried plums, unpeeled apples, pomegranates, artichokes, pecans, strawberries, green tea, and spinach.

*About <u>alcohol</u>: A little imbibing (one drink/day for women, two drinks/day for men) has been linked with lowering the risk of diabetes, kidney stones, heart disease, stroke, and cancer (for certain beverages). Alcohol in excess, however, causes liver damage, impaired memory and brain functions, anemia, heart disorders, depression, cancer (especially breast cancer for women), and many other health issues. Be very careful with your consumption and definitely do not binge drink since this can cause brain damage, heart attack, musculoskeletal disorders, Korsakoff's syndrome, and impaired judgments that can harm others as well as yourself.

<u>Sodium</u> is an important mineral for the control of fluids in the body and helps regulate the heart and metabolism, but too much can force the heart to work harder, so limit sodium intake to less than 2,300 mg/day (roughly 1 teaspoon). (Minimize processed foods!) <u>Caffeine</u> consumption affects people in very different ways. It helps some people boost their dopamine and norepinephrine levels, which can increase mental clarity, mood, and concentration in some, but in others it causes loss of sleep, lower concentration levels, muscle spasms, and heart arrhythmias. It has been linked to decreasing type 2 diabetes and might be beneficial to fight endometrial cancer, Alzheimer's, and stroke, as well as provide protection against sun damage. But some research suggests that caffeine can also cause breast cysts, bladder cancer, migraines, and sleep problems, so limit consumption of caffeine to 200–300 mg per day (~2 cups of coffee/day).

Sugar used in sweets, drinks, and processed foods should be limited to ~32 grams/day (8 teaspoons). (Sugars in fruits and dairy are not included in this number. Consuming the recommended 2–3 cups of dairy and 2 cups of fruit per day would be well over this amount, but luckily those are good sugars that our body is primarily using for energy rather than storing as fat.) Good "**brain foods**" include oatmeal (high fiber and grain is good for memory) and eggs; these have all essential amino acids and choline, which is good for memory. Flaxseeds are a good source of omega-3 fats for cell production. The five "worst foods" include microwave popcorn, hamburgers, foods with high fructose corn syrup, canned tuna, and canned tomatoes. (Bummer—I eat them all!)

Now that you know what to eat, you should also monitor how many calories you need to consume per day for your particular body size, age, and activity levels. It's important not to consume more than you need, so that you don't store too much fat. You can calculate your estimated body fat based on your weight and height with a simple equation:

Calculate Your Body Mass Index (BMI):

$$\frac{\text{Weight (pounds)} \times 700}{\text{Height (inches)}^2} = \text{BMI} \qquad \text{Example:} \quad \frac{132 \times 700}{67^2} = 92,400/4,489 = \textbf{20.6}$$

BMI is just a rough estimate of body fat. It doesn't take into consideration height, age, muscular makeup, or bone density.

Body frames (small, medium, large) will affect your BMI numbers. You can measure your wrist to figure out your frame size: <http://www.healthstatus.com/calculate/fsz>.

If you have dense bones, the BMI number will be higher.

There is a myth that muscle weighs more than fat, though 1 pound of fat weighs the same as one pound of muscle. However, fat takes up more space than muscle, so a person looks bigger if they are composed of more fat.

A person with more muscle will generally burn calories faster than a person with more fat, which you should keep in mind when you calculate your BMR (below).

For an average person (with average muscle and bone density) these are the guidelines relating to BMI:

> < 18.5 underweight
> 18.5–24.9 normal/healthy
> > 25.0–29.9 overweight
> > 30–32 obese (34% of the population in the U.S. is obese).

If BMI is > 27, your health risk rises dramatically, usually due to unbalanced "energy equation" (not burning enough calories for the amount you're taking in). Food energy is measured in kilocalories = 1,000 calories of heat energy; calorie is shorthand for kilocalorie. Okay, once you have your BMI you can see generally whether you need to increase your food consumption if you are underweight, keep it relatively the same if your BMI is in the normal range, or cut back on calories if you are overweight. **Then calculate how many calories you need a day. Basal Metabolic Rate (BMR)** is the amount of energy needed to maintain basic bodily functions. Active muscular individuals can eat more because they are burning more calories with the activities they are involved in as well as the fact that muscle tissue burns more calories than fat tissue does. As we age, we don't need as many calories per day, as the metabolism slows and activity levels start to drop off. So the equation to figure out how many calories should be consumed per day takes into consideration your weight, age, and activity levels.

Basal Metabolic Rate:

Weight (pounds) x 10 if sedentary	Subtract	0 if 18–34 = BMR
15 if moderately active		100 if 35–44
20 if highly active		200 if 45–54
		300 if 55–64
		400 if over 64

For a moderately active 20-year-old: 132 x 15 – 0 = **1,980 kcal/day**. (This example is more in line with a woman's weight/height; for a 20-year-old man, the average is probably around **2,550–3,000 kcal/day**.)

If your BMI was under or over the healthy range, then pick a weight for yourself that would put you in the normal range, and use that weight to calculate your BMR so that you know roughly how many calories you should consume for that weight. To lose weight it's best to not only cut back on calories but also to exercise regularly. You need at least 60–90 minutes a day of exercise along with a healthy diet for successful weight loss. It is best to start slowly, and don't go on the crazy fad diets, but do reduce your intake while increasing your activity levels. All types of exercise are great for building up muscles (which will speed up your metabolism), but aerobic exercise tends to shed the pounds faster. Exercising in the morning (before breakfast) has also been associated with faster weight loss. Save the strength training for the afternoon and don't exercise too close to bedtime since it will make it more difficult to sleep.

If you are maintaining a healthy weight it is still important to get at least 30 minutes of aerobic exercise 5 times a week and 20 minutes of strength training 2 times a week. This will help prevent weight gain and keep your muscles and therefore your metabolism healthy. And remember all of those other benefits of exercise from Chapter 6? I'm going to list them here again as a reminder.

Benefits of exercise:
1. Keeps muscles active and healthy and increases their efficiency.
2. Burns calories, and helps to maintain a healthy weight. You actually need less energy for metabolic functions if your body is toned versus if the body has excess fat.
3. Improves muscle coordination.
4. Develops new blood vessels and increases the amount of **myoglobin**. (Myoglobin is the pigment used to deliver oxygen in muscle tissue rather than hemoglobin.)
5. Develops more mitochondria in muscle cells (remember, this is where ATP is made).
6. Increasing the number of mitochondria in cells due to exercise will also increase your energy levels and the release of heat, important for thermoregulation.
7. Increases bone density and prevents bone loss.
8. Makes heart muscles stronger.
9. Pumps blood more efficiently, delivering oxygen and nutrients more efficiently.
10. Reduces blood pressure.
11. Reduces bad cholesterol and increases good cholesterol in blood (muscle contraction stimulates the release of lipoprotein lipase, which metabolizes triglycerides and helps to make HDL—the good cholesterol).
12. Better digestion and elimination of wastes.
13. Having healthy, strong muscles will give you a faster metabolism.
14. Reduces stress—very important, since stress weakens your immune system.
15. Helps with sleep—critical for health, growth, mental clarity, memory, etc.
16. Improves mood and increases alertness by stimulating the release of endorphins, serotonin, and/or dopamine.
17. Very important for your brain! People who exercise have less brain shrinkage during their life, and exercise enhances cognitive flexibility and function. (Studies show that kids who get to exercise regularly during the day perform better in school.) It also makes people more receptive to learning.
18. Increases your chances of a longer healthier life!

Fitness levels correlate to higher academic levels because exercise helps the brain function, delays cognitive diseases such as Alzheimer's, and in kids it increases attention and they perform more efficiently. So be sure to get a good night's sleep, eat a balanced diet, and get plenty of exercise!

Hopefully you have now calculated your BMI and BMR. Another great exercise to help you be more aware of how your diet corresponds to the nutritional recommendations is to record everything you eat in a typical day. Then tally your calories, and grams of carbohydrates (sugars and fibers), fats (saturated and unsaturated), and proteins. You can get all of this information from the labels on your foods or if you are eating fruits, vegetables, meats, dairy, etc., that do not have labels, you can look up the nutritional content on-line with different websites such as <http://nutritiondataself.com>. Once you have your totals, calculate your percentages for the day ... are you getting roughly 75% carbohydrates, and a combination of fats and proteins to make up the remaining 25%? What is your sugar intake versus your dietary fiber? How much saturated versus unsaturated fat are you consuming in an average day? This may be an eye-opening experience. Now look at the total amount of calories consumed and compare it to your BMR that you just calculated—are you consuming more or fewer calories than you should? You can also track how many calories you are expending per day and see if your BMR calculation is a good estimate of how much you should be consuming. There are websites that can help you figure out how many calories you are expending for different activities based on your weight and how long the activity is performed. <http://www.caloriesperhour.com/tutorial_BMR.php>.

Some regular metabolic activities and calories expended for a 132-lb person:

Sleeping: 60 kcal/hr
Sitting: 75 kcal/hr
Eating: 115 kcal/hr
Dressing: 100 kcal/hr
Computer: 120 kcal/hr
Standing: 140 kcal/hr

Versus some aerobic activities:
Walking: 285 kcal/hr
Moderate activity like hiking or biking: 356.4 kcal/hr
More strenuous activity like kickboxing: 594 kcal/hr
Running a 9-minute mile: 670 kcal/hr

(Remember, these numbers will vary based on weight)

Now let's see what this person—okay, me—would have to do to lose a pound of fat. If we assume that I sleep eight hours (480 kcal), and that eating, dressing, working on the computer, teaching, getting to and from school on my bike, and walking to get the kids from school is probably around another 1,500 kcal, I can break even here if I just consume 1,980 kcal for that day. If I want to LOSE weight, however, I would need to do a lot more. **Losing a pound of fat requires expending 3,500 kcal!** So if I consume 1,980 kcal, I'd still need to use up another 3,500 kcal! Looking at some of the numbers for the aerobic activities, I would need to walk for two hours, bike for three hours, kickbox for two hours, and then jog for an hour! I'm exhausted just thinking about that. Sounds like my Mother's Day last year when my daughters decided we should do everything I love in one day, so we played soccer, hiked, swam, biked, played tennis, and danced! I'm sure I lost a pound that day, but it's much easier to just not eat more than you need so that you don't end up needing to lose weight.

Okay, I will also give you some examples of how many calories are in some of the foods I love and how much activity I would need to expend to burn off the calories:

> 2 slices of pizza, 800 calories
>> play basketball for 1.5 hours
> 1 brownie, 280 calories
>> swim for 45 minutes
> 1 cinnamon bun, 170 calories
>> play soccer for 20 minutes
> 1 banana, 112 calories
>> check my emails for an hour.
> 1 orange, 65 calories
>> walk for 15 minutes or sleep for an hour

(You can do this for some of the foods you like too, to see how long it's taking you to burn things off. I've been typing at least an hour—I'm ready for another banana!)

Genes and activity levels affect weight as well as hormones, which influence **appetite** (pleasant anticipation of food we expect to enjoy), and **hunger** (when the brain tells the body it needs food), and how the body stores fat. Hormones travel in the blood to the brain. The region that controls hunger is the hypothalamus.

Four Hormones that affect hunger/appetite (just a reminder of the hormones we covered during the digestion chapter—also remember that it takes a while for these hormones to trigger the response, so eat slowly so that you will know when you are full before overstuffing yourself):

1. Stomach secretes **Ghrelin**, which is an appetite stimulator, when the stomach is empty. Ghrelin is secreted from stomach walls; when it enters the bloodstream (from small intestine absorption into capillaries), it travels up to the brain—the central area in the brain to receive these signals is the hypothalamus—which sends the signal to the body to eat.

2. Pancreas secretes **Insulin** when your blood sugar level is high and the hormone suppresses your appetite as it is working to spur cells to take up sugars from blood to be used for cellular respiration. "Sugar rush" occurs when insulin levels spike; then we feel lethargic as the cells absorb the sugar and insulin levels decrease. Insulin indirectly causes hunger by reducing blood sugar levels. When insulin levels are low, you burn fat, and when levels are high, you store fat. The more fat you have, the more insulin is pumped out per meal, so you end up storing more fat when you have more fat (a vicious cycle). Constant carbohydrate snacking causes insulin levels to go up and down so much that the body becomes desensitized to a certain degree and needs more insulin to get the job done. This insulin resistance is called syndrome

X. This raises triglyceride levels and can lead to type 2 diabetes, heart disease, metabolic syndrome, and obesity.

3. Small intestine secretes **PYY**: If you have enough food in small intestine, this hormone <u>suppresses</u> your desire to eat.

4. In the liver, your fat-storing cells secrete **leptin**, which <u>suppresses</u> appetite as the amount of fat storage rises.

When you lose weight, leptin levels fall, and you start to feel hungry again.

Some people are not producing or responding to some of these hormones properly and thus have a hard time regulating their appetite and hunger. This can lead to weight problems. Sometimes it's genetic and sometimes these hormone problems can be triggered by lifestyle choices. For example, lack of sleep can actually change these appetite hormones and cause weight gain! Getting fewer than 7 hours of sleep stimulates higher ghrelin and lower levels of leptin; therefore less sleep means increased hunger, which usually leads to higher BMI.

Sleep is very important for many reasons. The nervous system requires eight hours of sleep to repair neurons, the immune system is stronger when we get enough sleep, we process what we learn during the day during sleeping hours, muscles need time to recuperate for the activities of the next day, and other hormones (like the growth hormone) are released during sleep (that's why it's especially important for growing kids to get enough sleep). Kids need 9–11 hours of sleep, teens need 8.5–9.5 hours, and adults should get at least 7 hours/day. Studies have shown that people who get fewer than 6 hours of sleep a night have a higher risk of heart attack and stroke, so be sure to get a good night's sleep—it will also help you with your studies!

Nutritional Problems

Obesity is a huge problem in the Unites States, with nearly 300,000 deaths from obesity-related health problems every year. (It is one of the top causes of death in the United States.) Obesity can lead to high cholesterol, hypertension, heart disease, cancer, asthma, and type 2 diabetes. **Type 2 diabetes** more than doubles the risk of heart disease and also increases the chances of stroke, kidney disease, vision loss, dementia, sexual dysfunction, limb amputations, and shorter life expectancies.

Roughly 68% of adults in America are overweight or obese. Based on BMI statistics in America, 35.7% of adults and 17% of youths (ages 2-19) are obese, according to the Centers for Disease Control and Prevention. Obesity has tripled since 1980 and the problem is expected to get worse. It is a modern problem (within the last 50 years), largely the result of the preponderance of high fructose corn syrup, fast food restaurants and convenience foods, larger servings, **sugar drinks**, labor-saving devices, more sedentary jobs, motorized transportation, and less time for physical activity. This health issue is not only affecting millions of lives, but it is costing our country billions of dollars (73 billion in lost productivity and 150 billion in health care costs per year).

I bolded the sugar drinks because according to Robert Lustig, MD and colleagues at UCSF researching obesity, the amounts of high fructose corn syrup and sucrose we consume, especially in sugar drinks (juices, sodas, sports drinks, coffees, etc.), are a huge contributing factor to our higher obesity levels today. Americans drink 44.7 gallons of soda per year on average. This means that the average American consumes the equivalent of more than a can of soda per day. There are roughly 10 teaspoons of sugar per 12-ounce can of soda, which already exceeds our nutritional recommendation for sugar (we should only consume 6-9 teaspoons of sweeteners per day for a 2000 calorie diet.) This excess sugar is really helping to pack on the pounds. (A person drinking one can of soda per day, for a year, translates into 15-20 additional pounds if everything else remained constant.) The prevalence of added sweeteners in drinks also explains why some babies are obese before they even start eating solid foods—formula contains high fructose corn syrup and sucrose. Obesity in children is on the rise as well. Twenty three MILLION children in the U.S. are obese or overweight.

Basically, one third of all American children are overweight or obese! This is due to bad diets, too much TV and computer time, and not enough exercise (education budget cuts have caused a decrease in the amount of PE children get these days as well). Seventy percent of obese children will become obese adults. America is one of the leading nations in obesity. The current recession will likely cause even more obesity as people turn to cheaper foods as well (dollar menus at fast food places are a lot cheaper than buying fruits and vegetables at the grocery store, unfortunately). Lower socioeconomic status is linked to obesity for this reason as well.

People store fat in different ways
    1. "apple fat"—fat is above the waistline like a "beer belly" (more in men), which is more deadly in terms of heart disease.
    2. "pear fat"—fat is mostly on hips and thighs (more frequently found in women).

On the opposite end from obesity is **malnourishment** due to eating disorders such as anorexia and bulimia, health problems caused by a lack of appetite or by not being able to afford or access enough food. Being underweight can be just as deadly as being overweight.

The eating disorders anorexia and bulimia are both psychological disorders caused by the obsession to be thin.

Anorexia involves not taking in enough nutrients. This disorder occurs mostly in teenaged girls (the average age of anorexics is 17, but it can also go on into the early 20s and beyond). It is not just a white women's disease, as previously perceived either. Other races and males are anorexic as well, especially males who take in too many proteins and not enough carbohydrates in an effort to build muscle mass. Anorexia can be hereditary, but there are so many genes that are involved in this disorder that it is hard to pinpoint which ones cause this obsession to be thin. People who have a family history of anorexia are twelve times more likely to develop it themselves. Some other triggers for anorexia include the media's constant display of images of underweight teenstars, movie stars, and models. It is a very dangerous disorder and causes many health problems, including death.

Bulimia involves purging after eating (vomiting or laxative abuse). Bulimics are often normal weight or sometimes overweight due to the bingeing that occurs before the purging. Bulimia leads to electrolyte imbalance, acid damage to the upper GI tract, tooth decay, and nerve damage. Both anorexia and bulimia can lead to muscle atrophy, osteoporosis, digestive system problems, heart and kidney issues, psychological problems, and a weakened immune system.

Binge eating without purging is another eating disorder, usually due to emotional stress, and can lead to obesity.

Without treatment, 20% of people with serious eating disorders die. If you think you may have an eating disorder or know of someone who might, please get help. They are hard to stop on your own. Luckily, eating disorder treatments are now covered by some insurances under President Obama's expanded health insurance plan for psychological disorders.

People can be underweight due to other causes as well. Digestive problems, depression, thyroid issues, cancer, and other diseases can cause loss of appetite. If you are losing weight and you don't know why, you should definitely go visit your doctor. Some common causes are hyperthyroidism, celiac disease, or malabsorption disorders.

Malnutrition due to not being able to afford or access enough nutrients is another reason for underweight individuals. This occurs in 925 million people around the world. Malnutrition can lead to vitamin and protein deficiencies, developmental problems, anemia, immunity disorders, and death. This is a very serious problem worldwide.

If the body lacks sufficient amounts of carbohydrates (the normal source for glucose needed to produce ATP), it mobilizes fat stores and converts fat into small molecules called ketones. As ketones are oxidized to produce ATP, the body may enter a metabolic state called ketosis, which can lead to acidosis (acidic blood), which, in turn, can lead to a coma or death. (The sweet smell of ketone breath can reveal ketosis.) This condition can also occur if too much insulin is administered to a diabetic.

Health Guide

Here is a recap of some of the numbers you should try to maintain to stay healthy:

Blood pressure: Less than 120/80 (above 140/90 indicates hypertension)

Pulse: The normal resting rate is 60–80 beats per minute regardless of gender or age. The best time to check your heart rate is in the morning before getting out of bed. If you are very active the normal rate can be lower (50–60 beats per minute) and if you're pregnant your pulse is generally higher.

Blood sugar: The normal range of glucose is between 60 and 99 mg/dL or less at a fasting level (higher than 126 indicates diabetes).

C-reactive protein (CRP) in blood should be less than 3.0 mg/L (this is a marker of low-grade inflammation that is linked to heart disease, stroke, diabetes, and some cancers).

Height: Monitor your height to see if you are shrinking. If you lose more than an inch and a half after age 21, it suggests bone loss (osteoporosis), and you should get a bone density test.

Cholesterol: Total number should be under 239, LDL under 129, HDL above 45.

Triglycerides: Normal range is 150–199 mg/dL (increased levels indicate risk for type 2 diabetes and heart disease). If levels are high it is important to (if it applies) quit smoking, exercise, have no more than one drink per day, and reduce high-cholesterol foods.

Thyroid-stimulating hormone (TSH): From 0.1–5.5 helps regulate metabolism, body temperature, and heart rate; affects skin, hair, muscle strength, mood, and mental functioning. High TSH indicates hypothyroidism (thyroid is not producing enough hormones).

BMI: Between 18.5 and 24.9. BMI outside of this range is linked to risk of premature death due to many different health problems associated with underweight and overweight individuals.

(Most of these are recommendations from Kaiser Permanente; they can vary depending on whom you talk to, but these are the general ranges.)

To stay healthy it is recommended that you exercise daily (150–240 minutes of aerobic exercise/week), plus some strength training. Consume 5–9 servings of fruit and veggies/day, eat a serving of fatty fish twice a week or take 3 g of fish-oil supplement/day, 75 mg vitamin C (can get in a glass of OJ/day), 1,000 mg of calcium, 1,000 IU of vitamin D, limit alcohol to 1 drink per day max, don't smoke, get 7–9 hours of sleep/night, and reduce stress levels. A recent study published in *Internal Medicine* states that bad lifestyle habits (not following these guidelines) can cause you to age faster and increase the chance of premature death by about 20%. The most common causes of death stated were heart disease and cancer—both related to unhealthy lifestyles.

Number of deaths in the U.S. for the ten leading causes of death:

Heart disease: 631,636
Cancer: 559,888
*Stroke (cerebrovascular diseases): 137,119
Chronic lower respiratory diseases: 124,583
Accidents (unintentional injuries): 121,599
Diabetes: 72,449
Alzheimer's disease: 72,432
Influenza and Pneumonia: 56,326
Nephritis, nephrotic syndrome, and nephrosis: 45,344
Septicemia: 34,234

These data are from 2006. The information is a little outdated, but I wanted to include it so that you have a general idea of how many people die per year from these main causes of death. <http://www.cdc.gov/nchs/fastats/lcod.htm>.

*Stroke has been moved to #4 since this study came out due to more prevention and awareness. People are now suffering from strokes at a younger age, however, due to poor diets and higher obesity levels.

Chapter 12

Urinary System

We've looked out how the gases and nutrients get into the blood and travel all around to be delivered to the cells. We also saw how the liver converts a lot of these nutrients into different forms and helps to purify the blood. The next important system to discuss is how the body maintains the homeostasis of the blood in terms of acidity, electrolytes, water content, nutrients and salts. The urinary system has the critical job of sorting the blood, reabsorbing the nutrients, electrolytes, salts and water that we need to keep, and excreting the rest in the form of urine. The chemical makeup of our blood-fluid tissue must stay more or less the same even when we are adding some things and excreting others, so the main function of the urinary system is to keep us in balance and get rid of the wastes.

Functions of the urinary system

Monitoring salts and electrolytes: Na^+, Ca^+, K^+, Cl^-, H^+. (An electrolyte is any substance containing free ions that make it electrically conductive.) These electrolytes are especially necessary for maintaining the proper pH of the blood and for nerve and muscle cells (to generate nerve impulses, which then activate muscle cells). We can produce some of these ourselves, but mostly we need to get them from our diet. Sodium chloride (NaCl) is a salt, which dissociates in water to create an electrolyte solution. Salts are important for regulating the water concentration in the body. When salt levels get too high, the brain will stimulate the feeling of thirst so that we remember to stay hydrated. Serious electrolyte imbalances can lead to dehydration or overhydration. Symptoms can include feelings of pins and needles from numbness of the skin, muscle spasms, and arrhythmias. The kidneys monitor the electrolyte levels with the help of hormones—antidiuretic hormone, aldosterone, and parathyroid hormone—and will either flush out excess electrolytes or conserve electrolytes depending on the body's needs. (If the body becomes dehydrated, such as after strenuous exercise or after being sick with vomiting or diarrhea, it is important to replenish both water and electrolytes. Sports drinks, like Gatorade, have both. You just don't want to get in the habit of drinking more than you need, however, because these drinks also contain a great deal of sugar.)

Excretes toxins from body: the liver breaks down proteins into the building-block amino acids and then converts these to ammonia (NH_3), which is a dangerous waste product. The ammonia is then converted to less-toxic urea. When the liver cells break down nucleic acids into nucleotides, it forms uric acid. (Uric acid sometimes builds up in joints, especially the big toe, causing gout.) Both of these waste products need to be excreted with urine. (Aquatic animals excrete ammonia directly into the water, and birds excrete a concentrated uric acid, but we need to convert the ammonia to less-toxic urea and then add the uric acid and water to the mix before we can excrete it.) Since water is lost in this process, it constantly needs to be replaced.

- Urine is 95% water, the rest is **urea** (toxic breakdown of amino acids in proteins), **uric acid** (toxic breakdown of nucleotides in nucleic acid), **salt, broken-down hormones**, and **red blood cells** (which release a pigment from hemoglobin called urochrome that gives urine its yellow color), and **bicarbonate** (HCO_3^-), **H^+**, and **creatinine**. (Creatinine is generated in skeletal muscles from the breakdown of creatinine phosphate, which can be used as an alternate energy source for muscle contractions.)
- Urine is very acidic and helps to flush bacteria out of the urinary tract as well.
- We can excrete small amounts of salts, urea, and uric acid from the skin (in sweat) as well as from the urinary system. We can also excrete small amounts of alcohol through the skin and from the lungs when we exhale (a breathalyzer can test alcohol levels by monitoring how much is being exhaled).

Monitors acid level in the body: Acids are generated during various cellular reactions.
- pH: percentage of Hydrogen (blood is approx. 7.4). The body must maintain this neutral pH. In order to do this, the kidneys must either excrete H^+ or return it to the bloodstream, depending on the conditions.
- pH scale: 1–6.9 is acidic/high H^+; neutral = 7; 7.1–14 is basic/high OH^-.
- Carbonic acid is also returned to the blood to help as a buffer.

Regulates water content in the body: Water is the most important fluid in the body. Every part of our body (cells, tissues, organs, plasma, tissue, fluids, etc.) contain mostly water. Even the brain is 65–75% water! Animals are around 65–90% water (depends on the animals—for instance, jellyfish have a higher percentage of water than a lizard). Plants are usually 90–99% water (with the large water vacuole filling most of their cells).
- For humans, just a few days without water can cause death. (Cells can't perform their metabolic functions without water and they shrivel and die.)
- Too much water is dangerous as well—cells can pop, since there is no cell wall to prevent this from occurring. Too much water can also cause high blood pressure, and excess stress on the heart and kidneys, which can lead to heart or kidney failure.)
- Hypothalamus in the brain monitors water–salt levels and it makes you feel thirsty when sodium levels in the blood are high (which means that water levels are low).

- We need to take in the same amount of water from our foods and drinks that we lose per day, plus we make water with our metabolic reactions like cellular respiration. A good rule of thumb is to take in roughly half of your body weight in ounces of water. (So if you are 120 pounds you drink 60 ounces of water (7.5 cups) and if you are 180 pounds you drink closer to 11 cups of water.)

Functions of Water in the Body

1. Most chemical reactions occur after substances have been dissolved in water.
2. Water is liquid at body temperature and carries substances along in blood plasma. It delivers nutrients and helps to flush out waste.
3. Our cells produce heat during chemical reactions, and water can absorb the heat and keep our body at a constant temperature. Heat is produced during metabolic reactions and is transferred to the blood and sweat. We cool off when blood vessels in the skin radiate the heat and when sweat evaporates off the skin. If blood vessels get too hot, signals from the hypothalamus make the vessels dilate. We can lose about two liters of water a day through sweat, urine, feces, and vapor from lung evaporation. (So water evaporates from both lungs and skin.)
4. Water makes up most of the cell's cytoplasm, a key component in cell processes.

Water is critical for all of our body's systems, especially the digestive, circulatory, urinary, nervous, integument, and immune systems.

There is such a thing as taking in too much water: **hyponatremia**, or water intoxication. This dilutes the salt concentration in the body and puts too much pressure on the cells, blood vessels, brain tissue, and kidneys. Symptoms depend on how much water is consumed, how quickly, and the condition of the person. It can cause nausea, headache (due to the brain swelling in the skull), confusion, seizures, and sometimes even death. (Don't ever drink two gallons of anything in one sitting!)

Organs of the Urinary System

There are four main organs of the urinary system: kidneys, ureter ducts, bladder, and urethra.

Components of the Urinary system

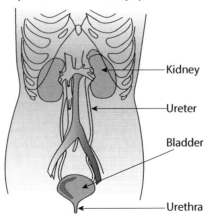

Kidneys are located in the upper part of the abdominal cavity, partially protected by the ribs, a little above the waist. They are only about 5 ounces each, but they use 20–25% of the body's oxygen and cardiac output despite their small size. The aorta brings blood to the renal arteries that enter the kidneys. The vessels get smaller, and finally the capillaries in the kidneys process the blood (filtration, reabsorption, and secretion occur), and then the renal veins bring the blood back out of the kidneys, where it will make its way back to the heart through the inferior vena cava. Kidneys cycle 45–50 gallons of fluid/day by processing our blood 30 times every day! Ninety-nine percent of the water and useful materials cycled through it are returned to the blood, and only ~1 liter of urine is produced and eliminated a day. Each kidney is made up of ~1 million **nephron** cells. These are the functional units of the kidneys. (This is more than the body actually needs, and therefore if one kidney is removed, the remaining kidney will provide the requirements for the entire body.)

Functions of the nephron cells:
1) Filter the blood
2) Reabsorb all the good nutrients, electrolytes, and water back into the blood
3) Secretion of larger particles and ions into the nephron to be excreted

- The kidneys monitor the amount of oxygen in the blood; if it is too low, the kidneys stimulate yellow marrow to convert to red marrow in order to make more red blood cells from stem cells in the red bone marrow.
- Kidneys release the hormone erythropoietin (EPO) to stimulate stem cells in bone marrow to make more blood cells.
- Kidneys also make the enzyme renin, which helps keep blood pressure within normal ranges.
- These organs not only purify the blood and form urine in the process, but they also adjust for water gains and losses.
- The kidneys help control the pH balance of the body either by excreting H^+/reabsorbing bicarbonate when the blood is too acidic, or by keeping the hydrogen ions and excreting the bicarbonate when the blood is too basic.
- They transform vitamin D into an active form with the help of the hormone calcitriol, which promotes growth and mineralization of bone by monitoring calcium levels.

Ureter is the tubule that leads from the kidneys to the bladder. It contains the urine formed in the collecting ducts of the nephrons. Smooth muscles in the kidneys as well as the ureter move the urine along by peristalsis.

Urinary bladder stores urine until it is ready to empty out of the urethra. The bladder usually holds roughly 1.5 liters of urine, but some people can store much more than this. The amount varies greatly. When the walls of the bladder (composed of smooth muscle) stretch to a certain point, stretch receptors signal to the central nervous system, which then triggers the need to urinate. There are two sphincters that need to open in order to release the urine from the bladder. One is under autonomic control and the other is under voluntary control (although if the bladder is too full, you can't really control this sphincter either).

Urethra is the tubule where the urine will exit from the body after being released by the bladder. The female urethra is only 2 in. long, males' are ~6 in. The male urethra travels in the penis and carries both urine and semen (although not at the same time), and the female urethra opens just above the vaginal opening, separate from the vaginal canal.

Now we will take a closer look inside the kidneys to see <u>how the nephrons work</u>:

Each nephron has multiple parts that are aligned in the kidney in a particular way. The nephron starts with a bulbous opening called the Bowman's capsule where one of the capillary beds enters. The Bowman's capsule then narrows to form the proximal tubule, which leads to the loop of Henle (descending and ascending limbs), and then the distal tubule finally joins with the collecting ducts that are all coming together to form the urine. The Bowman's capsules are located toward the outside of the kidneys in the cortex area. The loop of Henle descends into the middle portion of the kidney, referred to as the medulla, and finally the collecting ducts all converge at the area where the ureter is leaving the kidney, called the renal pelvis. Each portion of the nephron and therefore the kidney is responsible for different functions. Remember that there are a million nephrons in each kidney, and also remember that it is filled with capillary beds. The capillaries that enter the Bowman's capsules for filtration are called the glomerulus capillaries, and the capillaries that are responsible for reabsorption and secretion and that wrap all around the nephrons are called the peritubular capillaries. Now let's look at the different processes occurring all along the way. Don't worry if you don't get it the first time around reading this—it's very confusing. We will go over it slowly in class.

Here is my depiction of a **nephron** with the functions of each part listed below:

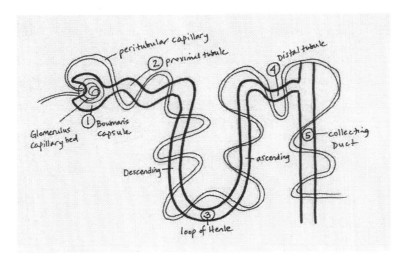

Nephrons control composition of blood by three processes:
- **Filtration**: Blood from the renal artery to the arterioles force fluid from the glomerulus capillaries into the lumen of the nephron tubule at the Bowman's capsule. Capillary walls are filters—they let out water and small solutes (salts, nitrogenous wastes, amino acids, vitamins, etc.) but not blood cells or plasma proteins.
- **Reabsorption** occurs at various regions along the nephrons. Some substances are reabsorbed and returned to the blood (sugar, amino acids, electrolytes/salts, water, and other nutrients). They make

their way from the nephrons to the peritubular capillaries, which then drain into venuoles, and finally out the renal vein.

- **Secretion** is very important for maintaining the blood's pH level of ~7.4. Many unwanted compounds in the blood are too large to be filtered in the tiny glomerulus capillary bed. Although the liver works to break down many of these undesirable compounds into small water-soluble molecules that can be filtered at the glomerulus, some products have to be secreted from the blood from peritubular capillaries into the tubular filtrate. This requires active transport using ATP. One of the main molecules secreted into the filtrate is H^+. The tubular membrane is not permeable to H^+, so once it is in the filtrate, it stays there. Remember how in respiration we saw how CO_2 in blood joins with H_2O to make carbonic acid, which splits to make bicarbonate and H^+? The bicarbonate ends up leaving the tubule to be reabsorbed into the blood (travels to lungs where CO_2 is expelled), but the H^+ is excreted out the nephrons in the urine. Other small molecules are also secreted into nephron tubules by active transport, such as potassium ions and ammonium ions, and drugs like penicillin, cocaine, marijuana, and other chemicals used for pesticides that we ingest in food. Secretion can occur from the blood to the nephron tubule at the proximal or distal tubules of the nephron as well as in the collecting ducts.

1. **Glomerulus/Bowman's capsule**: Filtration; blood is pushed out of the capillaries into the nephron tubule.
2. **Proximal Tubule**: Diffusion, facilitative diffusion, and active transport by cells lining tubule.
   a. Reabsorption occurs here and depends on active transport of sodium ions (Na^+) out of the tubule first and into the capillary bed nearby (the peritubular capillary), which attracts Cl^- to follow.
   b. Seventy-five percent sugar (by facilitated diffusion), water (osmosis), and amino acids are returned to the blood (reabsorbed into the peritubular capillary bed, which winds around the nephron).
   c. Fluid volume recovered into capillary/circulatory system.
3. **Loop of Henle**: Essential for concentrating the urine.
   a. <u>Descending limb</u> is permeable to H_2O, so water flows out by osmosis here. The peritubular capillary reabsorbs the water and the filtrate in the nephron becomes more concentrated.
   b. <u>Ascending limb</u> transports salts out of the tubule (passive transport near the bottom of the limb, and then active transport with protein pumps toward the top of the limb). The salts/ions exit out of the nephron and back into the blood. The ascending limb is not permeable to water, so no reabsorption of water occurs here.
   c. **Aldosterone** is a hormone that increases the reabsorption of sodium (Na) by activating sodium pumps in the upper part of the ascending loop of Henle and distal tube to facilitate salt (NaCl) recovery. Once salt leaves the tubule, the filtrate becomes more dilute, which makes the water flow out of the tubule as well by osmosis (in the distal portion). This hormone is released from the adrenal glands in response to signals from the brain about whether or not to conserve salt and water. It also triggers the release (secretion) of potassium in the kidneys. The action of this hormone increases blood volume and, therefore, increases blood pressure.
4. **Distal Tubule**: More salt and water reabsorption occurs here (controlled by aldosterone).
5. **Collecting Duct**: Permeable to water but not salt, so since the fluid outside the collecting duct is hypertonic (because salt was just pumped out at the loop of Henle and distal tubule), the water from the collecting duct will flow out by osmosis into the interstitial fluids around the collecting ducts. The degree of water recovery is hormonally regulated by **anti-diuretic hormone (ADH)**.

a. ADH facilitates H_2O recovery by promoting the opening of H_2O channels in the collecting duct.

b. Alcohol and caffeine inhibit ADH, so you don't recover H_2O in the body as much (this is why you need to pee a lot when you drink beer or coffee, so remember to drink plenty of water as well).

If you have plenty of water in your system, the ADH levels will be low, channels won't be opened as much, and water remains in the collecting ducts and therefore is excreted out with the urine. (After the filtrate leaves the collecting duct, it is finally called urine.)

As we age, our ADH levels go down and we are unable to retain as much fluid. This can cause frequent urination in older individuals.

Summing this all up:

Each nephron is serviced by an arteriole that subdivides to form the capillaries of the glomerulus.

Capillaries converge as they leave Bowman's capsule and form a second capillary network around the tubule called the peritubular capillary.

These capillary beds form into renal venuoles and then into the renal veins to take the blood out of the kidneys.

- The two capillary beds make up the **Portal System**.
- First capillary bed is where filtration occurs.
- Second capillary bed is where reabsorption and secretion occur. The secretion of H^+ and other ions are pumped into the nephron from the peritubular capillary for excretion.
- Nutrients are reabsorbed into the blood from the proximal tubule into the peritubular capillary as well as from the loop of Henle (where water and salts are reabsorbed), and the distal tubule (mostly NaCl), and then the collecting duct—more water is reabsorbed into the interstitial fluids.
- Nephrons all orient perpendicular to the kidney surface.
- Cortex contains Bowman's capsules and tubules, and medulla loops extend downward into the medulla and collecting ducts funnel into the pelvis area of the kidney.
- Aldosterone and ADH both play critical functions in salt and water balance.

Excretion refers to getting the urine out. Urine travels from the collecting ducts to the kidney's pelvis area, into the ureter, forced down by tiny urethral peristalsis (wavelike contractions), until it reaches the bladder. When the bladder is full, the stretching of the walls sends a message to the brain, which triggers the **micturition reflex** to empty the bladder (signal allows the sphincters around the urethra to relax). At the same time, the sphincters where the ureter meets the kidneys tighten so that the urine doesn't end up going back to the kidneys. Most people urinate 4–6 times a day. (Very light yellow urine signifies that you are getting enough water in your diet and so are able to excrete a fair amount in the urine. Very concentrated yellow urine is a sign that you need to drink more water. Cloudy pee can signal you are fighting off an infection since cloudiness means more WBCs in the urine.) Problems with the kidneys or other portions of the urinary tract can alter your regular urination routine. For example, urinary tract infections (UTIs) (which can occur anywhere along the urinary tract), make you pee more often, and cause a lot of pain. UTIs can be caused by bacteria, viruses, fungi, and parasites.

Urinary System Disorders

Kidney stones: Uric acid and calcium salts build up in kidneys (in the renal pelvis) and if large deposits can't get out with urine, they need to be surgically removed. If they are not removed and block the urine from exiting, they can cause kidney damage. Surgery to remove kidney stones involves crushing them with sound waves. Drinking plenty of water can help avoid kidney stones.

Polycystic kidney disease: An inherited disease, which forms cysts in kidneys and can lead to tissue death and kidney failure requiring dialysis or transplant.

Urinary cancer: Usually in the bladder or kidney. The cancer can easily spread from here to other parts of the body through the blood.

Urinary tract infection (UTI): Usually caused by bacteria. If it travels up the urethra into the bladder, it's called cystis or a bladder infection and can cause frequent and painful urination. This infection is treated with antibiotics, which kill the bad bacteria (remember to take the full course of antibiotics). Unfortunately, there is often a side effect of curing a UTI: a yeast infection! There are also beneficial bacteria in the urinary tract. Lactobacillus lives in the vaginal mucosa and produces lactate, causing a low pH, which helps keep other bacteria and fungi in check. Antibiotics kill both the infection-causing bacteria as well as the good bacteria. When this happens, the yeast flourishes, causing a yeast infection (very irritating). Often people take acidophilus (found in yogurts or supplements) to add bacteria back to the body to control the fungus (or treat medically to control the yeast levels until the body can rebuild its bacterial colonies). UTIs and yeast infections are more common in females because the urethra is shorter in women and more likely to become infected. Many women get more urinary tract infections when they become sexually active. If this is the case be sure to urinate soon after intercourse in order to flush out the urethra and always wipe from the front to the back.

Pyelonephritis: A urinary tract infection that has traveled all the way up to the kidney. Dangerous since this can lead to kidney failure if it is not treated in time.

Anuria: When urine can't be produced. This signals that there is an extreme health problem, possibly renal failure or complete urinary tract obstruction by kidney stones or tumors. Kidney problems (if severe) are often treated with **dialysis**, which can restore the proper balance of solutes, water, etc., by either carrying the blood from the patient and filtering it outside of the body, or pumping the proper dialysis solution into the patient's abdominal cavity; wastes diffuse into the tubes carrying the dialysis solution and get drained out. Some kidney disorders can be treated; for those that cannot, the patient either receives a new kidney or the patient dies.

Nephritis and **glomerulonephritis**: Inflammations of the nephron tubules that can lead to kidney failure. Symptoms include nausea, memory loss, dizziness, and fatigue. If the ion balance remains out of whack, death can follow.

Gout is from excess uric acid, which then crystallizes in a joint, usually near the big toe. (Can be caused by producing too much uric acid or not eliminating it efficiently.)

Kidney failure can occur when nephrons are not working any more. Nephrons can deteriorate due to diabetes mellitus, high blood pressure, infections, or toxins that poison the renal tubules. When the kidney(s) can't filter any more, the wastes will accumulate in the blood, blood pH becomes very acidic, and the electrolyte levels are out of balance. Acute failure can lead to convulsions, coma, and death.

Kidney transplantation can be performed if kidney problems are severe. Donors can be dead or alive (there are usually no adverse reactions to live donors if the extraction is done properly since we really only need one kidney to perform all of our urinary system functions.) The recipient needs to take immunosuppressants so that the new kidney is not rejected and it's best if the donor and recipient have compatible blood and tissue types.

Dialysis helps with kidney failure by filtering the blood outside of the body with a machine or diverting the blood to be filtered by a membrane in the abdomen. More than 200,000 people a year are on some sort of dialysis in the United States today.

Urinalysis helps to check what is in the urine: pathogens and white blood cells signal an infection, albumin signals hypertension, hemoglobin can mean there is bleeding in the upper urinary tract, excess glucose can signify diabetes mellitus, different hormones can signal pregnancy, steroids are checked for in athletes, drugs and alcohol can be detected, and blood in the urine might mean kidney stones are present.

Humans generally develop more urinary problems as we age since the kidneys get smaller and the blood flow to the kidneys decreases. (From age 40 to 70 the blood flow to the kidneys slows by 50%!)

Chapter 13

Nervous System

There are two ways our body's cells can communicate with one another: electrical signals from the nervous system (can communicate within milliseconds!) and the slower chemical signals from the endocrine system (hormones). These two systems are very interrelated, since the nervous system coordinates the endocrine system. In fact, the nervous system coordinates and controls just about everything our body does: voluntary movements, involuntary organ processes, reflexes, sensory system, mental processes (learning, analytical functions, language, thought), memory, emotions, dreams, thermoregulation, circadian rhythms, etc., etc.! The nervous system coordinates your body's responses to internal or external changes in the environment by causing a response in muscles, glands, and skin. This system works incredibly quickly to generate responses and works closely with the endocrine and muscular systems. The human nervous system is the most complex of all the animals, with <u>billions</u> of nervous system cells involved. (Some organisms don't even have a head or brain.) There are only two classes of cells in the nervous system: ten percent are **neurons** and ninety percent are accessory **glial cells** (aka neuroglia). The incredible neurons receive, process, and react to information and trigger a response. The glial cells have many functions to help the neurons perform these tasks.

Three Main Functions of Nervous System

1. Receive internal and external signals all over the body. This is done by the sensory system, which is part of the **Peripheral Nervous System** (PNS).
2. Process the information in the brain (unless it is a reflex, then it is processed in the spinal cord). The brain and spinal cord make up the **Central Nervous System** (CNS).
3. Cause a response by triggering muscle, gland, or skin cells (this occurs in the motor system of the Peripheral Nervous System).

Three main types of neurons perform these different functions:

1. **Sensory neurons** pick up stimuli from receptors (inside or outside of the body) in the PNS sensory system.

      <u>Three types of sensory receptors to pick up stimuli</u>

   1. Special sense receptors—associated with eyes, ears, nose, taste buds
   2. General receptors all along skin—sense temperature, pressure, pain
   3. Visceral receptors—within our body; monitor organ function and help to maintain homeostasis (stomachache and sore throats are felt with these receptors)

   Information from the stimuli is then sent for processing in the spinal cord (for reflexes) or the brain via the interneurons.

2. **Interneurons** make up 95% of neurons in the nervous system (we have approximately 100 billion!). Interneurons are mostly found in the brain and spinal cord (CNS)
   Interneurons are between sensory and motor neurons

3. **Motor Neurons** are attached to muscle, gland, or skin cells and cause a response, which can be a voluntary or involuntary depending on which part of the motor system is involved:

Divisions of the Motor System

Somatic: Mostly voluntary (under the conscious control of skeletal muscles). Somatic responses can also have to do with reflexes that are not under voluntary control, but still involve skeletal muscles. All somatic neurons release the neurotransmitter acetylcholine (ACh), which causes skeletal muscles to contract.

Autonomic: (unconscious control, involuntary) Smooth and cardiac muscles are involved in these responses. There are two autonomic systems in the PNS:

<u>Parasympathetic</u> responses: Normal state, digestion, respiration, energy storage, relaxation.
- Thick saliva produced, decrease in heart rate, increased production of digestive enzymes, speeds up food movement, homeostatic functions.
- Involves nerves only in the cranial, cervical, and sacral regions.

<u>Sympathetic</u> responses: Excited state [four Fs]: Fight, fright, flight, and fornicate.
- The neurotransmitters and stress hormones epinephrine and norepinephrine increase for sympathetic responses. (They are released from the adrenal gland as hormones and from the brain as neurotransmitters.)
- Dilation of pupils, thin saliva produced, dilation of bronchial tubes, increased heart rate, digestion slows down, goose bumps, sweat, stimulation of sex organs.
- Sympathetic responses only involve nerves in the thoracic and lumbar regions of the spinal cord.

Now let's look at the parts of a neuron cell:

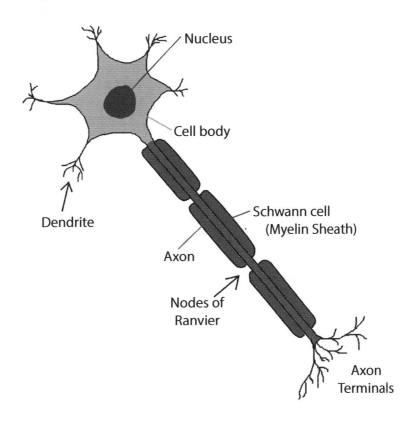

Starting from left to right: the **dendrites** have a very large surface area to receive the signals from stimuli from receptors and pass it along to the cell body and then the axon. The cell body with the nucleus is called the "soma." The **axon** moves the impulse, in the form of an <u>action potential</u> (AP), from the start of the axon (at the <u>axon hillock</u>), down to the end of the axon, which terminates in the axon buds, and then the impulse is passed along to the next neuron via neurotransmitters. The neurotransmitters are released from the axon's terminal buds into the synapse (the space between the neurons or neuron and muscle/gland). There are neuroreceptors on the dendrites to receive the signals from the neurotransmitters. Depending on which neurotransmitter is received determines what the next neuron will do with the information.

- The end of the axon can have multiple branches (terminal buds) so that one axon can transfer the signal to many neurons, muscles, or glands (depending on what kind of neuron it is). (A neuron may have as many as 10,000 synapses with other neurons!)
- The axons in our arms and legs can be incredibly long. For example, the sciatic nerves are composed of our longest neurons due to these elongated axons. They are our longest nerves and radiate from our hips to our knees and then branch further as they continue on to our feet. Most neurons and nerves, however, are much smaller—especially the neurons in the brain.
- In most neurons there is a myelin sheath covering the axon (made from Schwann cells in the peripheral nervous system). These Schwann cells are a type of glial cell. They are in units like penne pasta

with Nodes of Ranvier between the units. They provide insulation around the axon that helps the nerve impulse move faster by allowing the action potential to jump from node to node.

- <u>Functions of glial cells</u>: Provide metabolic functions for neurons, regulate ion concentrations around neurons, and help with insulation (which makes nerve impulses travel one hundred times faster!). And while they are wrapped around axons, they can help axons regenerate themselves if they are damaged. (This only occurs in the PNS by <u>Schwann cells</u>. There are other glial cells in the CNS called <u>oligodendricytes</u> that provide insulation for those axons but do <u>not</u> have this regenerative function. Therefore axons that are damaged in the CNS cannot be repaired.) They also supply neurons with nerve growth factor, without which neurons would die.
- Neurons traveling into the spinal cord (into CNS) are called <u>afferent neurons</u> and neurons coming back from the CNS to the motor neurons are called <u>efferent neurons</u>.

Nerves: Bundles of axons. They can be from motor or sensory neurons or both types together.

Ganglia: Bundles of cell bodies and dendrites. (The "basal ganglia" refer to the ganglia in the brain. Ganglia group interconnected nuclei within the cerebral cortex, thalamus, and brainstem; associated with a variety of functions, such as motor control, cognition, emotions, and learning.)

As I've mentioned, the space between neurons is called a **synapse**. Therefore when referring to neurons on either side of the synapse, they are called the presynaptic and postsynaptic neurons. These are all of the parts of neurons, and now we will look at how they work:

How Neurons Work (This is a VERY basic description, but this is just an overview and I don't expect you to know the intricate details of neuronal conduction.)

Neurons are excitable and react to different electronegativity. Stimuli have to be strong enough to cause a response. The response is either all or nothing, and the strength of the nervous stimulation varies depending on how many neurons are firing at the same time. A nerve impulse is sent down an axon, starting at the axon hillock, due to changes in the charge across the neuron membrane, called the **neurolemma**. Protein channels inside the neuron are triggered to open if a stimulus is strong enough. **Sodium (Na+) channels open first**, and sodium goes into the membrane. This changes the balance of ions, which causes an **action potential**. Then **potassium (K+) channels open** to release potassium outside of the neuron. The action potential sends the message down the axon in a wavelike movement. Action potentials move down axons fastest if the axons are covered in thick myelin sheaths (fatty glial tissue) and the action potential can jump from one Node of Ranvier to the next.

There are sodium-potassium pumps that will pump sodium back out of the cell and potassium back into the cell with the use of ATP, and the ions will go back to the original concentrations on each side of the neurolemma. Once the neuron is back to the resting potential (after the action potential is complete), it is ready for a new action potential if another stimulus arrives. The action potential (nerve impulse) moves down the axon and causes the release of neurotransmitters from the presynaptic neuron to stimulate receptors on the postsynaptic cell (which can be a neuron, muscle, or gland) to cause a response such as another action potential, a muscle contraction, or a gland secretion. (They can be stimulating or inhibiting neurotransmitters.) For example, the motor neurons around muscles release the neurotransmitter acetylcholine (ACh), which triggers Ca+ to help contract muscles. Neurotransmitters are in vesicles and can be broken down by enzymes from postsynaptic neurons (like ACh), or can get reabsorbed by presynaptic neurons to be used again after it has caused a response (like norepinephrine).

So if we follow a <u>stimulus</u> all the way through the circuit, it might look something like this: a finger is near a flame, the <u>general sensory receptor</u> on the finger collects the information about the heat of the flame and this signal is picked up by the <u>sensory neuron</u> in the PNS. The action potential from the sensory neuron gets relayed by neurotransmitters to <u>interneurons</u> in the spinal cord, traveling all the way up to the brain. The brain processes that the finger should be removed from the heat, and the message is sent back down through more interneurons back to the PNS and finally to a <u>motor neuron</u>, which then releases the neurotransmitter actetylcholine (ACh), which triggers the muscle in the finger or hand to move away from the flame. (If the heat was so intense that it caused a reflex response, then the message wouldn't even need to go all the way to the brain, since reflex responses come directly from the spinal cord.)

Neurotransmitters are incredibly important chemicals in our body. The type of signal received will determine which neurotransmitter is released, which, in turn, will cause an effect on the neuron, muscle, or gland that the neurotransmitter is acting upon. We all have different levels of each of the neurotransmitters depending on genetics, activities, diet, and experiences. They help make us who we are, how our bodies function, how we feel, and how we act. We are used to a certain baseline amount of these chemicals and will often engage in activities or crave certain foods in order to get ourselves back to our own balanced baseline. Some people also alter their chemistry by using alcohol, drugs, caffeine, or medications—all of which can alter the functions of neurotransmitters by creating changes at the synapse where the neurotransmitters are released. There are more than one hundred neurotransmitters, but I'm just going to go through the top eight here and explain what they are related to (in most people). I will also mention some ways to boost some of these chemicals in your brain and body.

Main Neurotransmitters:

Acetylcholine (ACh) is found all over the peripheral nervous system, as well as in the CNS. It is our most prevalent neurotransmitter. In the brain it is important for memory and learning, and also affects our mood since it is an inhibitor of dopamine. In the peripheral nervous system, it helps with muscle contraction by triggering the sarcoplasmic reticulum to release calcium in the sarcomere. (Foods that increase ACh levels are things with choline, such as eggs and soybeans.)

Norepinephrine (in brain) is responsible for excited rush (adrenaline rush) in tense situations, also important for temperature control, attention, and consciousness. It is important for regulating mood in the pleasure center, and essential to hunger, thirst, and sex drive. (Norepinephrine can be a neurotransmitter when it is in the brain and it can also be a stress hormone when it is released from the adrenal glands.) It is an excitatory neurotransmitter that speeds up thoughts and blood pressure (similar to dopamine, which is the precursor to norepinephrine). Protein foods, competitive sports, and stressful, scary, or sexual situations and activities will increase this chemical.

Epinephrine has effects similar to norepinephrine but is found in different parts of the brain. Both of these neurotransmitters are hard on the body in large amounts, since they will speed up your heart rate and increase anxiety. Epinephrine also dilates airways (that is why it is used if an allergic reaction constricts airways), and slows down the digestive system. (Epinephrine can be a neurotransmitter when it is in the brain and it can also be a stress hormone when it is released from the adrenal glands.)

Dopamine—(in brain) is also an excitatory neurotransmitter and is the building block for norepinephrine. It speeds you up (makes you think, speak, and breathe faster), helps regulate emotional responses, functions with ACh to help muscles contract smoothly, and is also responsible for pleasurable experiences. Because of this, trying to attain high levels of dopamine often leads to addictive behaviors. People will do things (like obsessively clean, exercise, do drugs, work or gamble, etc.) because the activity raises their dopamine levels and they feel like they need to maintain these levels in order to feel good. Falling in love raises dopamine levels as well.

- Cocaine and amphetamines increase production of dopamine as well as inhibit the removal or reuptake of dopamine from the synapse so that it continues to activate receptors for an abnormal period. This can lead to bursts of energy, delusions, paranoia, and euphoria. It will also lead to depleted dopamine levels in the future, which causes depression, lethargy, muscular problems, and sometimes suicide. Cocaine abuse weakens the immune system in general and can raise the risk of heart disease since it makes the heart work at a faster pace than normal.
- People with naturally excessive amounts of dopamine often exhibit heightened activity and emotions, and even hallucinations and paranoia (schizophrenics may have increased levels of dopamine).
- Aerobic exercise can deplete available dopamine and trigger the body to produce more. The more you use, the more you make. Protein-rich foods will also increase dopamine levels. If you feel your levels are low and need a boost, 3 to 4 ounces of meats, fish, beans, eggs, etc., can raise your levels within 10 to 30 minutes!

Serotonin—Another feel-good neurotransmitter, but this one is more of a relaxed, content, joyful, peaceful chemical. It is important for emotional state, memory, mood, sensory perception, and body temperature. Increased serotonin elevates mood, self-confidence, euphoria, and restful sleep, and reduces sensitivity to pain.

- The drug Ecstasy causes neurons in the brain to release too much serotonin, which results in elevated moods while using the drug, but after time it uses up the body's natural supply of serotonin and can lead to depression. Overdoses can be very dangerous as well, since they cause high blood pressure, heart failure, high body temperatures (which cause enzymes to denature—lose their shape and function), and sometimes death.
- Antidepressants, like Prozac, also work to elevate your mood, but they work in a more regulated way in order to not deplete your serotonin levels over time. They work by blocking the reuptake of serotonin into the presynaptic neuron so that it stimulates the postsynaptic neurons longer. (There are also antidepressants that are dopamine and norepinephrine reuptake inhibitors.)
- To boost serotonin levels naturally, you can do mild aerobic exercise (walking, biking, yoga, stretching, etc.) or even just a relaxing activity that usually boosts your mood such as reading, meditating, watching TV (nothing too stimulating or graphic—that will boost the dopamine level), or listening to your favorite music!
- Certain foods can also raise serotonin levels. The amino acid tryptophan is a building block of serotonin. Foods high in tryptophan will help produce more serotonin (although tryptophan in meats doesn't work as well since it's harder for the amino acid to get to the brain when it is packaged with that amount of proteins and fats). Carbohydrates are the best source to boost serotonin

levels—especially whole grains (brown rice, oats, wheat), flour products (bread, bagels, pasta, rolls, crackers, cakes, cookies, etc.) and starchy vegetables like potatoes and yams.

Histamine is a neurotransmitter in the hypothalamus, which regulates sexual arousal, pain, thirst, and blood pressure.

Endorphins are produced and released from the hypothalamus and pituitary gland, especially during strenuous exercise, pain and injury, excitement, and sexual arousal (orgasm). They are our natural opiate and produce a feeling of well-being as well as function to relieve pain. They can inhibit neurons from releasing "substance P," which is a pain indicator. (Painkillers like Oxycontin ["hillbilly heroin"] do this as well and are very addictive. This is now one of the top abused prescription drugs.)

GABA—inhibits other neurotransmitters and therefore slows down responses to stimuli. This neurotransmitter helps to lessen anxiety and relieves tension.
- Yoga and other types of relaxing exercises can increase GABA levels and promote a more relaxed state with less anxiety.
- Alcohol and Valium both increase GABA levels. The "drunk" feeling from alcohol is when this neurotransmitter is inhibiting other neurotransmitters or responses, and especially affects neurons in the cerebellum (which helps with motor control and balance) as well as in the cerebral cortex (affecting decision-making and perceptions).
- Studies have found that many alcoholic fathers have sons with lower levels of GABA. This in turn may cause some of the sons to become alcoholics as well in order to decrease anxiety, since they don't have as much of the natural coping mechanism.
- Excessive repeated abuse of alcohol can cause nervous system problems (such as degeneration of neurons), along with liver failure and heart damage, so if you drink, be sure to do it only in moderation.

One more chemical that is secreted from neurons that you may find interesting is nitric oxide (NO). This is actually secreted as a gas from neurons and causes blood vessels to dilate. An example of how this gas works is penile erection. A stimulus will cause nitric oxide to be released from neurons in the brain, which will trigger the blood vessels to dilate in the penis. Blood rushes into the vessels and an erection occurs.

Let's move on to nerves, the spinal cord, and the brain.

There are thirty-one **Nerves** coming into/out of the spinal cord or brain (each nerve has sensory and motor neurons at each end):
twelve pairs Cranial Nerves, right from brain
eight pairs Cervical Nerves
twelve pairs Thoracic Nerves
five pairs Lumbar Nerves
five pairs Sacral Nerves (*The sciatic is the longest nerve in the body.)
one pair Tailbone/Coccyx

The **Spinal Cord** is a highway for carrying signals between the PNS and the brain. Nerves are threaded through a canal in the vertebral column and ligaments attached to bone. The spinal cord is filled with cerebrospinal fluid and has three layers of membrane tissue called <u>meninges</u> around it. The cerebrospinal fluid and meninges provide protection and help with nourishing the CNS. The spinal cord is the control center for <u>reflexes</u> (autonomic reflexes like controlling the autonomic sphincter for emptying the bladder).

The **Brain**, I think, is the most awesome structure in our whole entire body! All of our behavior, feelings, thoughts, organ system functions, movement, etc., are coordinated by this three-pound structure! It is composed of roughly 100 billion interneurons and is aided and protected by cerebrospinal fluid, special membranes called meninges, and then our skull and scalp.

Here is a simplified diagram of the brain:

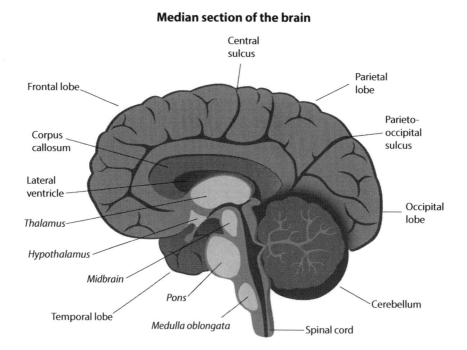

Median section of the brain

Now let's go over some of the main components of the brain and its protective coverings and fluids.

The brain is protected by the **scalp** (where the hair is attached), and then the **skull** bones (cranium) under the scalp, followed by the meninges layers. Three layers of **meninges** cover the brain. From the outside in, there is the 1) dura mater, 2) arachnoid mater, and 3) pia mater (with cerebrospinal fluid between #2 and #3).

The **cerebrospinal fluid** (CSF) is constantly produced and absorbed into the blood. The four large cavities in the brain called <u>ventricles</u> each contain a <u>choroids plexus</u>, which forms the CSF. It is transported from the ventricles to the center of the canal in the spinal cord, and then back up the spinal cord (on the outside of the canal), to the meninges, where it is absorbed into the bloodstream in the arachnoid layer. It helps to absorb shock, support and cushion the brain, as well as carries out the circulatory functions of delivering nutrients, ions, oxygen, hormones, red and white blood cells, and water to the cells of the brain (the neurons and glial cells). It also carries away the waste (like CO_2), which gets transported from the brain capillaries to the venuoles and finally back into the superior vena cava heading into the heart. The capillaries in the brain are very selectively permeable (the CSF helps to maintain the blood–brain barrier) and they allow gases, water, and nutrients, but do not allow large molecules like antibiotics to move into the CSF (that's why it's hard to treat infections in the brain). The capillaries do let lipid soluble substances into the brain like caffeine, nicotine, alcohol, heroin, etc., and anesthetics. These substances pass through the lipid plasma membranes of the cells in the capillary walls to the neurons, where they cause an effect on brain function very quickly. (Capillary beds in other parts of the body have pores between the walls' cells that nutrients, gases, etc., can pass through, as well as larger substances. The capillary beds in the brain do not have these pores, and the cells are very tightly packed together to reinforce the barrier even more.)

There are four main parts of the brain:

1. **Cerebrum** (forebrain): The largest part of the brain and is the highly developed information-processing center. This area is important for how we learn, remember, plan, analyze, etc. Appropriate responses are generated here and nuclei are connected from here to all other parts of the brain. The very thin outer layer of the cerebrum is called the **cerebral cortex**, which is the conscious part of the brain composed of tens of millions of neurons. (It is only about 4 mm thick, however!) The cerebral cortex area is organized into four different lobes with motor and sensory strips in between two of the lobes (further descriptions to follow). The lobes include tissue from the cerebral cortex area as well as the area that extends down from the cerebral cortex called the "**associated areas**." When we are conscious, the lobes are constantly collecting and processing input from touch, sight, sound, body position, muscle movements, etc. The portion of the interneurons in the cerebral cortex is known as the "gray matter" since it is composed mostly of cell bodies (with nuclei) as opposed to the "white matter" that extends inward, composed of the myelinated axons (the fatty myelin sheaths appear white). The order is reversed in the spinal cord: the gray matter is toward the inside and the white matter is toward the outside. (These "gray" and "white" references are antiquated descriptions from old staining processes, but I like them since I read a lot of Agatha Christie mysteries growing up, and the detective, Hercule Poirot, was always referring to his "little gray cells" that would help him solve each case.)

   The **cerebral cortex** is responsible for taking in amazing amounts of information for conscious activity, while the **associated areas** in between the outer layer and the limbic system in the core are important for interpreting, integrating, learning, speaking, remembering, and acting on sensory information and memories. The cerebrum is receiving countless messages constantly. It processes the messages, sends the information to appropriate parts of the brain, and causes reactions all of the time within milliseconds. It is the brain's ultimate conscious control center. As we move toward the center of the brain, on the other hand, we find the limbic center, important for unconscious automatic responses, emotions, and memory.

2. **Diencephalon** (midbrain—where the limbic system is housed): Includes the **thalamus** (two egg-shaped structures); important for sensory processing from taste, sight, touch, and sound. (It receives information

from all of the senses other than smell, which goes to the olfactory cortex.) The thalamus then forwards the sensory message to the area that will process it in the cerebrum. (The different lobes are associated with different senses). The **hypothalamus** (just below it) controls the homeostatic environment. It has both neurons and hormones, so it is involved in both electrical and chemical signaling. It regulates water balance, smooth muscle contraction, body temperature (keeps it ~98.6 °F), circadian rhythm, appetite and thirst; the center for pain and pleasurable reward and sexual behavior (by triggering the release of hormones involved in the reproductive system), it secretes six releasing and inhibiting hormones that control the anterior pituitary gland (the "master gland" that secretes seven hormones that work on different systems all over the body). It also makes antidiuretic hormone and oxytocin, which are stored and released from the posterior pituitary gland. The pituitary is right below the hypothalamus, which we will discuss more in the hormone chapter. The midbrain is home to the limbic system, which is the emotional and memory center as well, and includes the **corpus callosum** (connects the right and left hemispheres, sharing information and generating appropriate responses), **hippocampus, amygdala, cingulated gyrus** (areas for memory, emotions, and gut responses), the olfactory cortex and the **olfactory bulbs** (where the nerves bring in signals from the nose). The limbic system will be discussed in more detail below.

Some other structures in the midbrain:

Reticulate Formation is a network of fingerlike nerves coming from the brain stem between the ears into the thalamus to help relay signals from the spinal cord as well as to help control arousal.

The **optic chiasmata** is where the nerves from the eyes cross as they head to the back of the brain to the occipital lobe. Each eye has a right and left field of vision. The information from the left half of each field goes to the right side of the brain and the information from the right half of each field of vision goes to the left side of the brain. Data from each side will then cross back to the opposite side through the corpus callosum as the two hemispheres communicate to process the information. There is also an area near here called the **suprachiasmatic nucleus**, which acts as the master biological clock of the body. Light received from the eyes will trigger this area to help control our circadian rhythm (cycles that help us anticipate when there will be food, light, and heat available). This helps us to know when to sleep, eat, and heat up. (As morning approaches, the light will trigger this area to stop the production of melatonin, raise the body temperature, and feel hungry. Our temperature peaks during the day when we are most alert and our thinking and memory are the sharpest, and then falls around siesta time and drops even more at bedtime. (Aha! This is why I always have on multiple layers by 9 pm!)

Pineal Gland releases the hormone melatonin to help regulate sleep cycles.
- Late-afternoon light collected from the eye receptors and relayed to the suprachiasmatic nucleus will help trigger the production of melatonin from the pineal gland. There is a delay in the release so that it occurs in the dark, and the production will be suppressed by sunlight in the morning. People who travel east should make sure to get some afternoon light at the end of the day to help their bodies get back on track with producing melatonin to reduce jet-lag effects.

- In some areas where there is not enough sunlight (e.g., Alaska), or in the winter, increased levels of melatonin will consume serotonin (our relaxed, feel-good neurotransmitter) and people often feel depressed (SAD = **S**easonal **A**ffective **D**isorder).
- Melatonin production decreases at puberty and may help trigger gonads to start developing more. It decreases as we age even more, making it more difficult to sleep for long stretches as we used to be able to do when we were young (if you are still able to "sleep in," enjoy it now).
- We already discussed the **benefits of sleep** in terms of weight and growth. (The pituitary gland releases growth hormone during sleep.) It's also incredibly important for processing the information we receive throughout the day and for storing memories. It helps with critical thinking, the immune system (helps the body recover from tissue damage and helps eliminate free radicals that are produced during respiration in high amounts when we are conscious). And as we have all witnessed firsthand, sleep helps with mood, creativity, concentration, communication, and performance the following day.
  - **Tips for a good night's sleep**: Limit alcohol and caffeine 4–6 hours before bed, establish a calming bedtime routine and turn off all electronics, don't eat a large meal or exercise right before bed but DO exercise earlier in the day and spend at least 30 minutes outside daily. If you need to nap to catch up on sleep, do it at least 3 hours before bedtime. Try to get at least 7 hours of sleep per night for adults (more for kids and teens).
- There are five stages of sleep every 90 minutes, which range from slow-wave sleep (easily awakened) to the deepest REM (Rapid Eye Movement) stage, where we dream. We don't usually remember our dreams unless we are awakened during this deep sleep, but during a lifetime we might have more than 100,000 dreams!

3. **Cerebellum**: (part of the hind brain) Controls movement (motor control), posture and balance, helps judge time, sounds, and textures, and stores its own kind of motor memories. This is because cerebellar learning stores "skill" memories such as riding a bike (once you have learned to ride a bike, you never forget how to do it). Another type of motor memory that helps me is the act of writing things down. (Notes can enhance cerebellar learning and help you memorize material as well. So even though you have all of this material in the book, it is still good to write it down in class—it will help you remember things for the tests. I always have to write things down multiple times before they are really stored in my brain!)

4. **Brain Stem**: (below midbrain, also part of hindbrain, in front of the cerebellum) Contains the **medulla oblongata** (responsible for blood pressure and digestion, and respiratory and digestive reflexes like sneezing, coughing, snoring, hiccupping, swallowing, and vomiting), and the **pons** (coordinates communications between the cerebellum and forebrain as well as respiratory reflexes such as triggering inhalation and exhalation). The brainstem connects the brain to the spinal cord, so it is the transit point for all information going into and out of the brain. This portion is the most protected in the central lower region of the brain, which makes sense as it is the most primitive in terms of evolutionary development. It governs our basic functions for survival and reproduction, whereas the outer areas of the brain have the more advanced tissue (not as developed in most animal species) governing the higher thought processes.

Now looking more in depth at some of the regions already covered:

Limbic System: Composed of a number of structures in the brain, all in midbrain region. It governs instinctual drives and behavior, such as eating, anger, sex, fear, and fighting, and helps with maintaining homeostasis. It also releases hormones and governs emotions, memory, learning, and sensory and motor information and processing. The structures include:

- **Hypothalamus and Thalamus**—see #2 above for all of the different functions.
- **Cingulated Gyrus** (two structures, one in each hemisphere)—governs emotions.
- **Hippocampus** (one branch on each side)—a very important region for memory formation, retention, and retrieval. People who generally have to remember a lot of information tend to have enlarged hippocampus areas—this also demonstrates that brain tissue can physiologically change depending on the brain functions. New research has found that we <u>can</u> grow new brain cells, mostly in the hippocampal region. Exercise (both physical and mental, using your brain to learn something new), increases this neurogenesis. Exercise helps to stop brain shrinkage as well (our brain tissue shrinks about 1% per year starting in our late 20s.)
- **Amygdala** (one on each side) helps to process emotional information, gut reactions, and memories.
- **Olfactory Cortex** to receive stimuli from olfactory bulbs.

Memory: This is incredibly important, so I want to discuss it a little further here. The function of memory is to be able to use information that you experienced or learned in the past to help guide your actions in the future. We are receiving an incredible amount of information all of the time and don't need to remember it all for our survival or daily functions, so we can filter out what we need to keep and what we need to store. If information is irrelevant, it is forgotten quickly; otherwise it is transferred to long-term memory storage.

- Fact-based memory: Numbers, names, odors, faces. Goes to the <u>limbic system</u> for storage (mostly the amygdala and hippocampus). We bring the memory back up to the cerebral cortex for processing during recall.
- Skilled-motor-activity memories (like riding a bike) go to a different memory circuit, all the way back to the cerebellum. Writing down your biology notes may help you remember the information, since you are consolidating the memory by the act of writing it down.
- Memory occurs in three phases:
  o <u>Immediate memory</u>: Keeps information about immediate details such as where you are in conscious part of brain (cerebral cortex).
  o <u>Short-term memory</u>: Easily forgotten when not needed but helps carry out daily tasks, keep conversations going, remember things for exams, etc.
  o <u>Long-term memory</u>: Memories that are stored in the limbic system and can last a long time, possibly your whole life if they are really strong memories. Sometimes they can be associated with a sense such as smell or taste that will bring the memory back (these are the only two senses that come directly toward the hippocampus where memories are stored). The cingulated gyrus governs the emotions that may be stirred due to the memory. Memories are often associated with something important that occurred in life, such as the day you graduated, got married, gave birth, first relationship, etc. The more often you recall these memories, the more likely they will stick around.

As mentioned earlier, the act of writing down information, events, etc., will help solidify memories since they will become part of your motor memories as well. Keeping a journal is a good way to help you remember life events. Rereading, discussing, thinking about, and studying all help to solidify memories as well. (So make sure you have some time to do all of this before your next exam.) Sleep also helps your memory, because your brain needs time to process and store all the information that it is taking in, so try not to pull an all-nighter before exams. The strong memories I have about my life are all from stories that have been told over and over, and now are passed on to my kids, who love these stories as well (especially if the stories involve them). So don't forget to talk about your memories with others, and in this digital age, looking at pictures can help a great deal too.

There are two forms of memory. Explicit deals with conscious, intentional memories, generally processed by the left hemisphere. (For example, what you use for recalling information for exams.) The other type is called implicit, which is processed by the right hemisphere. These memories have more to do with perceptions and sensations and tend to generate emotions and behaviors. Implicit memory is also involved in motor memories formed in the cerebellum.

Ten steps to a healthy brain and optimal memory according to the Harvard Medical School:
1. Exercise—healthier lungs, more oxygen to the brain, more neuron connections, and enhanced neurotransmitters. It helps with memory, learning, attention, problem-solving, and mood. Also helps to slow physical brain shrinkage.
2. Keep learning—challenge your brain. This keeps the brain cells stimulated and communicating with one another, which is important for brain function.
3. Don't smoke—increases the risk of stroke and hypertension, both of which impair memory.
4. Alcohol in moderation—moderate alcohol consumption can reduce the risk of dementia and stroke; too much is toxic to neurons and is the leading cause of Korsakoff's syndrome, which causes memory loss.
5. Maintain a healthy diet—fruits and veggies especially help brain function.
6. Vitamins if necessary—vitamin E may help or delay Alzheimer's disease. Calcium and vitamin D are also very important to take if you don't get enough in the diet.
7. Adequate sleep—this is when memories are consolidated and stored.
8. Be social—helps reduce stress, which damages the brain, and social interactions promote stimulating activities.
9. Protect brain—from impacts and toxins, and protect hearing and sight as well.
10. Manage stress—since it can lead to altered brain chemistry and damage to the hippocampus.

Let's talk a little more about the **Cerebrum** and how people react, think, behave, learn, etc., differently due to differences in brain development, hormones, and which hemisphere is dominant.
*   Recall that the brain has both right and left hemispheres and that each side has specific functions that it controls more than the other.
*   Corpus callosum helps the two hemispheres in the cerebrum to communicate with each other. Even though one side may be dominant for particular tasks, both sides can play a role, and the entire cerebrum is aware of the sensory input and motor output.

Left Hemisphere: Usually the dominant hemisphere for most people. It helps to coordinate speech, language interpretation, writing, calculation, and logical decisions; controls the right side of the body.

- Some characteristics of "left-brained people": organized, list-makers, think in words rather than images, advanced planners, approach problems in a linear fashion, may excel at math, writing, and problem solving.

Right Hemisphere: Very musical, artistic, good at analyzing sensory input, recognize spatial relationships and faces, emotional interpretation of conversations is highly attuned; the right hemisphere controls the left side of the body, so often left-handed people are right-brained (not always, however).

- "Right-brained people" tend to be visually oriented, often thinking in images rather than words. They may be very scattered but think in terms of the big picture rather than in details. They are often talented in the arts and spatial design.
- Three to five percent of the population is left-handed or ambidextrous (where the dominance is shared equally), although most of this population is more right-hemisphere dominant.

Sometimes the dominance of the hemispheres is easy to detect and sometimes people use both hemispheres more or less the same. My two daughters are like night and day, and I can see why when I think of which hemisphere is dominant for them. I definitely have one of each. Luckily for me, despite their differences (or maybe because of them), they get along extremely well.

In terms of differences in brain depending on gender, there is quite a bit of debate whether or not there are any conclusive structural differences in the brains of men and women. Male and female brains are very similar in size if you relate the mass of the brain proportionally to the rest of the body mass, but there is one area that has been proven to have structural differences. This area is called the preoptic area in the hypothalamus, which is related to mating behavior (it is two times larger in males). Some other theories about differences in male and female brains include how brains develop (hormones play a role in this), and how males and females may use different parts of their brains more than other parts.

MALES:

Use analytical center for problem solving, good at mathematical calculations. Often have better spatial skills (can visualize objects in 3D better). Male brains are slower to develop, but usually they catch up at puberty since the sex hormones have an impact on brain development. Decision-making and control of aggression is in the prefrontal cortex, which doesn't fully develop until the 20s. This may be the reason younger men are more aggressive and are more likely to take risks, although this is also the time when men have more testosterone, which also affects their aggression levels. (You don't usually see 50-year-old men at the skate park getting into fights with one another.)

FEMALES:

Use both hemispheres for problem solving. Females are generally more attuned to the limbic area of brain; use verbal cues and emotions to process information and can express emotions and artistic appreciation more clearly. The female brain is good at detailed planning and multitasking. Females develop quicker, so may outperform their male classmates until the male brain catches up at puberty. (This is important for teachers of elementary school children to realize, so they can adjust their expectations.)

Different lobes of the cerebrum have different general functions as well, and are associated with different senses:

Four Lobes of the Brain:

Frontal: This area is involved in voluntary movements, speech, some memory, making judgments, inhibiting inappropriate behavior, and the **olfactory** center is here. **Broca's area** in the frontal lobe's left hemisphere controls muscle movement involved in speech (movement of tongue, lips, and mouth). Damage to Broca's area would stop the person from actually forming the words but he/she would still be able to read and understand both written and spoken words, since there are other parts of language in the temporal and occipital lobes. If the frontal lobe is damaged due to concussion, people can lose some memory, sense of smell, sometimes speech, logical thought, and the ability to recognize consequences. What is affected usually depends on the location and severity of the concussion.

Parietal: Receives input from bodily sensations (**touch**) and body position. There are two strips of brain tissue that go across the top of the brain between the frontal and parietal lobes, which are called the **motor cortex** (helps control voluntary movement) and the **sensory cortex** (receives and processes the sensory input from touch and movements). There are different amounts of neurons receiving input from different areas of the body in this region—the more neurons there are for an area, the more sensitive the area is. There are a lot of neurons in the sensory cortex receiving stimuli from our mouth (lips and tongue) as well as genitals, so these areas are extremely sensitive.

Occipital: **Vision** (nerves from the eyes come all the way to the back of the brain). There is an important aspect to language in this area as well. Not only does the **visual cortex** allow us to see words on a page, but the **angular gyrus** transforms the words that we see when we are reading to an auditory code so that we are hearing the words in our head as we read. Damage to this area can stop a person from being able to read (although they could still speak and understand words that are spoken).

Temporal: **Hearing**, as well as auditory and visual processing. This area allows us to recognize faces as well as comprehend the language of others and of ourselves. If it is damaged, the comprehension of written or spoken language can be lost. (Both the frontal and temporal lobes are active during speech, since the frontal lobe is controlling the mechanical movements of the mouth, tongue, etc., and the temporal lobe is constantly checking for the accuracy of what you just said.) **Wernicke's area** is responsible for language comprehension and expression. If this area is damaged, a person would be able to speak and read but not comprehend the words.

Nervous System Disorders and Problems

There are MANY, so I will just touch on some of the most common problems that you've probably heard of. I was surprised to read that 60% of all illnesses are caused by psychological conditions!

Paralysis: Nerve damage along the spinal cord that affects movement and loss of sensations. (Remember, neurons cannot repair themselves in the CNS like the PNS neurons can.)

Headache: Feels like pain is inside the brain, but there are no pain receptors in the brain, so it's actually stretching of muscles or tension of blood vessels of the face, scalp, or neck. Pain stimulus is sent to the hypothalamus where we recognize pain.

Migraines: May be due to increased sensitivity to pain in different parts of the brain and this can be hereditary. Migraines are three times more prevalent in women than in men. They can last up to three days and are often triggered by stress, changes in hormones, diet issues, and weather changes. Symptoms include pulsing on one side of the brain, pain, nausea, and light sensitivity.

Concussion: Traumatic brain injury (TBI) is the most common brain injury and can vary greatly in severity and in the results. A blow to the head or neck can cause loss of consciousness, memory, speech, olfaction; blurred vision; or even paralysis. Repeated concussions can also lead to symptoms very similar to Alzheimer's. The dementia that results is due to a buildup in the brain of <u>tau protein</u>. Many professional football players and boxers develop dementia due to this <u>chronic traumatic encephalopathy</u> (CTE). Symptoms include behavioral and personality changes, disinhibition, and irritability, and then it progresses to dementia as more neurons die.

Acquired Brain Injury (ABI): Nervous tissue death can be due to insufficient oxygen, nutrients, or water getting to the brain, or neurotoxins causing degeneration. Causes of ABI can be stroke, aneurysm, heart attack, brain tumor, seizures, infection, substance abuse, and malnutrition.

Alzheimer's: Degeneration of neurons due to buildup of beta-amyloid and tau proteins, plaques, and tangles, all of which leads to memory loss, mood swings, confusion, poor judgment, dementia, and loss of intellectual functions. (Similar to CTE, but Alzheimer's has beta-amyloid proteins as well as tau proteins.) The areas that are most affected are the hippocampus (memory) and the cerebral cortex (intellectual functioning), where the neurons die due to protein plaque. Many of the neurons in these areas normally secrete acetylcholine. (Remember, this neurotransmitter is mostly important for memory, reasoning, and learning when released in the brain.) The ACh levels in Alzheimer's patients' brains are often 90% lower than normal! Some patients take ACh replacement therapy, but it doesn't always help. Early-onset Alzheimer's is genetically determined, but this is not as prevalent as the kind that is attributed to old age and inactivity of body and brain, abuse of toxins, etc. (See the list of ten ways to maintain a healthy brain and optimal memory to reduce the risk of getting Alzheimer's.) The number of people with Alzheimer's disease doubles every 5 years after the age of 65.

Hydrocephalus: "Water on the brain" is due to the cerebrospinal fluid building up pressure under the skull, causing the neurons of the cerebral cortex to be compressed, which can shut down parts of the brain. (This happens if the CSF circulation or drainage is blocked somewhere.) Surgically implanting a shunt to drain the fluid relieves this problem.

Epilepsy: Seizure disorder caused by brain injury, birth trauma, or genetics, when the brain's normal electrical activity becomes erratic.

Stroke: (already covered in Chapter 8.)

Korsakoff's Syndrome: Damage in the thalamus and memory centers due to vitamin deficiencies from malnutrition, usually from heavy alcohol abuse.

Parkinson's: Degenerative neuron disorder due to loss of dopamine-producing neurons in the thalamus region. It usually occurs after 50 but some people can show symptoms earlier as well. People with Parkinson's have trouble with motor coordination and often have jerky movements since they are not making enough dopamine and norepinephrine (neurotransmitters that are associated with acetylcholine in stimulating smooth voluntary muscular movements). Other symptoms include slowness of movement, impaired balance and coordination, depression, and other emotional changes.

Multiple Sclerosis: An autoimmune disorder thought to be due to genetic problems or sometimes triggered by a virus. The body attacks its own myelin sheaths around the axons, which affects nerve impulses and can lead to slurred speech, fatigue, vision problems, and muscle failure.

Huntington's Disease: A hereditary degenerative brain disorder caused by a dominant allele. People with Huntington's don't usually know that they have it (unless they get genetic testing) until they're in their 30s, when the neurons in the brain start to deteriorate, causing loss of voluntary control in areas such as speech and movement.

Stress: Caused by a number of things: relationship problems, school, job, money issues, the desire to succeed leading to anxiety about performance, etc. As mentioned above, the stress hormones/neurotransmitters epinephrine and norepinephrine are released more during sympathetic responses and can get you through a difficult situation, but chronic stress is not good for the heart (increased risk of heart attack if the heart rate is continually high); people with a lot of stress often have digestive problems, and immune systems are weakened, which leads to a higher risk of cancer. (So remember to relax when you can—exercise, yoga, dance, meditation, etc., are all good stress relievers and helpful for your body.)

> **Post-Traumatic Stress Disorder** (PTSD) is stress triggered by a traumatic event, and it seems to get worse rather than better with time. It can lead to panic, inability to concentrate, memory problems, and fear. PTSD usually affects the limbic system, particularly the amygdala and hypothalamus.

Attention Deficit Hyperactivity Disorder (ADHD): Leads to a hard time concentrating or focusing due to poor functioning in the <u>reticular activating system</u>, which maintains alertness. This system, centrally located in the brain, has concentrated nuclei to receive sensory input, which it then directs to other portions of the brain to cause appropriate responses. This is the most common behavior disorder for children 3–17 years old, affecting 9% of kids in this age range in the United States. It can also persist into adulthood, but the earlier a child is treated for it, the better the outcome usually. ADD and ADHD are both neurobiological disorders that run in families so they are hereditary.

Cancers of the Brain: Caused by abnormal glial cells replicating out of control (gliomas).

Amyotrophic Lateral Sclerosis (ALS, Lou Gehrig's Disease): Degeneration of motor neurons, which leads to muscle failure and eventually respiratory failure. It does not affect sensory or interneurons.

Infections of the Brain: Lead to inflammation and compression of neurons. Infections are hard to treat, because the blood–brain barrier is very selective and often does not let antibiotics pass from the capillaries in the brain to the cerebrospinal fluid to the area of infection.

Encephalitis: Due to viral infections such as herpes or West Nile viruses.

Meningitis: Bacterial or viral infection in meninges membranes.

Abnormal Chemistry in the Brain: Can lead to a variety of problems, and affects millions of Americans; one in six families has some sort of mental illness present. Some of the problems are genetic and affect the levels of neurotransmitters and how people react to them, and some are acquired for different reasons. Much is still unknown about these disorders.

Depression: Unipolar mood disorder causing a person to have a low mood—feelings of hopelessness, not enjoying activities that they used to, low self-esteem, digestion and sleep problems, thoughts of suicide (~3% of people with depression do commit suicide, the third-leading cause of death in teenagers), fatigue, and can lead to problems with work, school, relationships, etc. The neurotransmitters that are involved with moods—serotonin, dopamine, and norepinephrine—are often not balanced in people with depression. It can be hereditary in some regards or acquired. Nineteen million Americans are clinically depressed, but if they can get help, medication and psychotherapy are very effective.

Bipolar disorder: Thought to be hereditary and due to chemical imbalances in neurotransmitters. More than 5 million Americans are bipolar to varying degrees. Problems with neurotransmitters lead to mood swings from manic to depressed states and cause erratic behavior, instability, sleep problems, lack of self-control, irritability, and on the other end, extreme happiness and bursts of creative energy (many bipolar people are extremely talented and smart). Medication is prescribed, but it is very hard to regulate this disorder. Most people do not know they have it until late in their teens to early 20s, and many end up self-medicating, which can lead to substance-abuse issues as well. The number of children diagnosed with bipolar disorder has soared from 20,000 to

800,000 in the past 10 years! There tends to be a correlation with children who have more than 2 hours of screen time per day being more at risk for social and mental problems in general.

Schizophrenia: Also disrupts brain activity, causing paranoid delusions from auditory hallucinations. It is most likely due to neurotransmitter problems as well, but the causes of this disorder vary.

Autism: Also due to abnormal chemistry in the brain. Causes for this vary as well, but genetics may play a part, since autistic children often have some family history of nervous system disorders or chromosomal abnormalities. It is a developmental disorder of the brain and affects social and communication skills. Many autistic kids are very sensitive to touch, smell, hearing, etc., and have mood swings and irritability as well. Treatments include medication, occupational and physical therapy, and diet changes (some respond well to a gluten-free diet). Autism affects males three times more than females. Other related disorders include Asperger's and Rett syndromes.

Other mental disorders: Panic attacks, obsessive-compulsive, anxiety, phobias, eating disorders such as anorexia, bulimia, etc., etc.

Nerve Disorders are prevalent as well and many are due to vitamin B deficiences. Some common nerve disorders are:

Sciatica—pressure on the sciatic nerve as it exits the spinal cord, which can cause pain all the way down the leg to the foot in some people.

Bell's Palsy—paralysis of one side of the face (usually temporary).

Carpal Tunnel Syndrome—compression of nerves in the wrist.

(If you think you may have a mental illness, don't be embarrassed to seek help. It is so important to get proper care so that you can start to feel and function better. Many people turn to illegal drugs to help regulate their moods, and that only makes things worse because it damages the brain even further. If you have mental illness in your family, there are also ways to get help. There is an organization called the National Alliance on Mental Illness (NAMI) that helps family members cope with the problems of having loved ones suffering from these disorders. They offer support groups and workshops and put out a quarterly newsletter. In the last issue that I read, Dr. Jill Bolte Taylor, a renowned brain specialist, wrote an appeal for people with mental illnesses as well as their family members (with normal-functioning brains) to donate their brains after death. If you have the organ donation sticker on your license, it does not include your brain, but they desperately need more brains to study in order to try to understand and treat mental illness. The brain is harvested within 72 hours after death, and it doesn't cost anything to be a part of the program. If you are interested, you can contact the Harvard Brain Tissue Resource Center's Psychiatric Brain Collection department at 1-800-BRAIN-BANK.)

Speaking of illegal drugs, let's cover some of the basics:

<u>Mind-Altering Drugs</u>: Bind to neurotransmitters in the brain so that the brain receives altered messages.
- Many affect the pleasure center in the <u>hypothalamus</u>. Drugs also affect consciousness, behavior, motor control, sensory abilities, appetite, and can alter heart rate and respiration.
- The body slows down the natural production of neurotransmitters over time due to drug use, and therefore the body becomes dependent on the drug to continue to feel good and function. Eventually the body can build up a tolerance for the drug of choice, and it takes more and more of the drug to produce the desired effect.
- Habituation is when a drug user craves the drug; this signals addiction.

Stimulants: Nicotine, caffeine, cocaine, and amphetamines all raise the heart rate and blood pressure, and can stimulate the release of neurotransmitters like serotonin and dopamine to elevate mood and produce an intense alert feeling, surge of energy, etc.

Cocaine: Stimulates pleasure by blocking the reabsorption of dopamine (affects fine motor control and learning). Weakens immune and cardiovascular systems.

Alcohol: A depressant that affects levels of GABA neurotransmitters, which are inhibitors. Alcohol abuse not only deteriorates neuron function, but it is also toxic to the liver (and with barbiturates, alcohol can be fatal). But as I've mentioned before, alcohol is thought to have some beneficial qualities when consumed in moderation.

Morphine and **heroin**: Both block pain by binding with specific receptors in the CNS and can create euphoria. They are derived from poppies and are highly addictive. Prescription Oxycontin has the same effect on people and is very addictive as well.

Marijuana: Affects the brain by binding to and activating specialized chemical receptors in the hippocampus, cerebellum, and frontal cortex, which control memory, perception of pain and pleasure, concentration, and coordination. We have these receptors for a neurotransmitter that we produce ourselves called anandamide (discovered in 1992). Anandamide seems to help with forgetting certain things, immunity functions, pain and pleasure responses, as well as helping to regulate the blastocyst implanting in the uterus. Anandamide has a very similar shape to the active ingredient of cannabis, tetrahydrocannabinol (THC), which also will bind at these cannabinoid receptors. The results of THC binding at the cannabinoid receptors meant for anandamide are many, including affecting the production, release, and reuptake of dopamine and other neurotransmitters. Emotional swings, increased appetite, and memory loss from smoking pot are all due to the fact that it mainly affects the limbic system. High doses can lead to increased heart rate, panic attacks, and learning/attention disorders, coordination problems, and fertility issues. In some cases, it can possibly lead to schizophrenia (cannabis smokers are 40% more likely than non-users to suffer psychotic episodes). This drug affects people very differently, however, and some people use marijuana for medicinal purposes (to ease pain and relax muscles), and to help their attention and learning. Marijuana is not physically addictive, but can be mentally addictive, and research shows that people who smoke marijuana are more at risk of becoming addicted to other drugs. Marijuana does not cause lung cancer like tobacco does. Even though one joint is equivalent to 2.5–5 cigarettes in terms of smoke damage, tobacco is more damaging overall since it is smoked in much higher quantities and is much more addictive. However, marijuana alters brain chemistry and can deteriorate neurons and connections over time, leading to memory and cognitive problems.

Hallucinogens: Psychedelics (mushrooms, mescaline and LSD), dissociatives (nitrous oxide) and deliriants (LSD) all cause changes in perception and emotions. These drugs can alter consciousness so that the user has little control over his/her actions and is disconnected from reality. Some affect dopamine and serotonin systems and can affect memory, and cause depression or euphoria, mania, attention problems, sensory dissociation, amnesia, and loss of consciousness. They are dangerous because they can impair judgment, which can lead to injury of the user or those around them. They are, however, sometimes used for medicinal purposes; in fact there is a study using hallucinogens to relieve anxiety in terminally ill patients.

- More than 22 million Americans have some sort of substance abuse problems, whether from alcohol, drugs, or prescription medications; fewer than 10% of these people actively seek help.
- Addiction warning signs:
  - If a person has to take more of a substance to get the desired effect (tolerance).
  - User feels like they need the substance to function.
  - Can't seem to stop even if they want to, and start to hide substance abuse from others.
  - Often they become defensive if confronted about substance abuse and will engage in dangerous activities in order to obtain the substance.
  - Deteriorating work or school performance and relationship problems are also a major sign.

If you or someone you know has substance abuse problems, get help as soon as possible before it gets worse.

Bottom line: **Protect your brain!** You can't repair the damage once it's done, and the brain is one of the most important organs in your body. It governs your emotions, actions, thoughts, memories, reflexes, etc., etc., etc. It governs your entire body, so PLEASE be careful what you do to it.

Chapter 14

Sensory System

S enses are critical for survival. They help us to find food, reproduce and rear young, detect danger, and help the body maintain the homeostatic environment. The fine-tuning and diversity of our senses make life more interesting, too. Think if all food tasted the same, all sounds were the same pitch and tone, we saw only in black and white, or we couldn't tell the difference between smells. We have very complex sensory organs to help us detect the stimuli, and then our highly developed brain processes the information that we gain from the senses we detect. Let's just review a little of the information that we discussed in the last chapter in terms of the sensory system.

Sensory System: When a stimulus (sound, light, heat, touch, odor) activates receptors (in ears, eyes, skin, muscles, nose), the stimulus can generate a nerve impulse, which travels to the brain/spinal cord (CNS). The thalamus is the area that processes the sensory information and sends it on to other areas of the brain to cause a response. A **sensation** is the conscious awareness of a stimulus. This occurs in the cerebral cortex, which processes the information of the sensation and forms a <u>perception</u>, an understanding of the sensation. If the sensation stirs a memory and emotion, the limbic system gets involved again, since the memory may be stored in the hippocampus or amygdala, and the cingulated gyrus deals with the emotions that the memory or sensation cause. We already discussed the different parts of the nervous system and how signals travel around in there; now we're going to look at the very special sensory organs that allow us to receive the signals so that we can taste, see, smell, hear, and touch the world around us.

Think of how complex and important your senses are. Your roommate makes chocolate chip cookies, the special sense receptors in your nose detect the odor, and the stimulus causes a nerve impulse in the sensory nerves. The impulse is then transferred to interneurons, which travel directly to the olfactory bulbs in the brain (these nerves don't have to go to the spinal cord first). The impulse is passed along to the thalamus, which processes the stimulus and sends it to the cerebral cortex, where the sensation and perception are formed. The cortex helps you understand what this smell means and where the smell is coming from (cookies in the kitchen!) and might even remind you of your mom baking you cookies and give you a happy feeling (impulse may have traveled from the cerebral cortex to the limbic system, and then back up to the cortex). All of this occurs in less than the blink of an eye, and you might already be out of your chair to go get some cookies. If this is the case, the thought process telling you to get up (from your cerebral cortex) innervated your cerebellum (motor control center) and then sent nerve impulses down through the spinal cord, and the interneurons brought the message out to the motor neurons. Acetylcholine released from the motor neurons trigger calcium to be released and your muscles' cells contract. This causes your muscles to move your skeleton, and you make your way toward a cookie. This is really a super-simplified version of what is happening, but it is just to give you a glimpse (or a sense!) of how senses, working in conjunction with the nervous system, are amazing.

There Are Three Types of Sensory Receptors:

General receptors are on the outside of the body, all along the skin. They can sense pressure, pain, temperature, and pleasure. The fingertips and the tongue have the most of these receptors.

Visceral receptors are on the inside of the body, on the walls of the organs, muscles, etc. Some of the nerves coming from these receptors just go to the spinal cord (which can cause a reflex reaction), and others continue on to the brain.

Special receptors are associated with the special sense organs—eyes, ears, and nose. These messages always get sent directly to the brain. (The nerves don't go through the spinal cord; they come directly from the outside of the head to the brain.)

KIND OF RECEPTORS:

Chemoreceptors: Odor, taste (nose, mouth)
Mechanoreceptors: Pressure (ears, skin, etc.)
Thermoreceptors: Heat (skin)
Pain receptors: Nociceptor (several million of these, everywhere except the brain)
Photoreceptors: Light (eye)

Chemical Senses:

There are roughly one hundred thousand chemoreceptors (for taste) on the tongue in the papillae bumps, on the roof of the mouth, and in the back of the throat. Different areas are specialized for detecting the five different types of tastes: bitter, sour, salty, sweet, and umami (which is savory like aged cheese or meats). Each taste bud contains all of the different taste receptors, but they have different concentrations of each receptor type. For example, there are more of the sweet receptors in the taste buds at the tip of the tongue, and more bitter receptors at the back of the tongue. (Bitter is our most sensitive receptor; we can sense toxic chemicals in very low doses, probably an evolutionary adaptation our primitive ancestors had when sampling plants to eat.) The sour and sweet receptors are more concentrated on the sides of the tongue, versus the umami, which is located more at the back of the throat and the roof of the mouth. New research is looking into whether or not we also have a taste receptor for fat.

Olfactory Receptors:

Humans have 10 million chemoreceptors in the nose. They are very specialized for different odors. (A bloodhound has 4 billion!) Chemoreceptors detect chemicals that travel into the nose and stimulate the receptors that are underneath the mucus there. The odor stimulates nerves that go directly from the nose to the olfactory bulbs in the brain; the limbic system is also involved, since odors often stir up memories and emotions. Chemoreceptors in the nose also play a role in our sensitivity to taste. The odors from the chewed food travel into nasal passages, and the perception from the brain helps to enhance the flavor of the food. (Notice that when you have a cold and your nose is stuffed up, food doesn't taste quite as robust.) We can detect direction of smell as well (signals from each nostril go to different parts of the brain and help us smell "in stereo" too).

There are also receptors to detect **pheromones**, signaling chemicals given off with body odor and sweat, that are important for social interactions. Humans don't use these as much as some other animal species (to coordinate whom to mate with and when), but compatible pheromones may have something to do with "love at first sight" and helping you decide whom to date. Research shows that pheromones may be detecting immunity "fitness," possibly enabling us to pick mates based on the best immunity genes. They are also the reason many women living together will often menstruate at the same time.

Hearing Receptors:

Very sensitive <u>mechanoreceptors</u> (sensory hair cells) are located inside the ears. Hearing starts when compressed air (sound waves) gets funneled from the outer ear ("pinna" made from cartilage) into the auditory canal, and the pressure vibrates the eardrum, which is the start of the middle ear. The eardrum, also known as the tympanic membrane, passes vibrations to three small bones in the middle ear: **mallus**, **incus**, and **stapes**. It's like a domino effect of one bone triggering the next to move, and then finally the stapes triggers the next stage of the process inside the inner ear.

The stapes causes the inner ear oval window (a membrane like the tympanic) to move the fluid in the **cochlea**. The mechanoreceptor hair cells are in the cochlea. When the tips bend due to the movement, it causes neurotransmitters to be released and a nerve impulse is sent to the brain to be interpreted as sound. The frequency of sound waves affects the pitch, and how many receptors are stimulated at once will determine how loud the brain perceives the sound to be. If these receptor hair cells are damaged for some reason (one can be born with the damage or acquire it from exposure to repeated loud noises), hearing will be impaired or lost completely. Sometimes it is caused by the sensory neurons not detecting the signal properly, and sometimes it is caused by the message not being sent or processed by the brain properly. It's very hard to correct in either case.

Another type of hearing loss can be due to sound not being conducted from the outer ear to the inner ear properly. This is called conductive deafness; hearing aids can increase the amplitude of the sound in order to help with this problem. Children have very sensitive hearing and can sometimes hear a pitch that adults can't hear any more because their receptor hairs haven't been damaged yet. As we age, our hairs become damaged and it becomes harder to hear. (With our iPod craze these days, one way to lessen the damage to our receptors is to keep the volume level at the middle level or lower.)

Tinnitus is a condition that causes a constant ringing in the ear sensation and can be due to an infection, hardened ear wax, a benign tumor on the acoustic nerve, or other problems in the ear. There is usually no treatment, but sometimes it will go away on its own over time. Avoid alcohol, nicotine, and loud noises because those tend to make it worse.

Some receptors in the ear also monitor equilibrium, so the ear is very important for balance and detecting gravity. Motion sickness is due to overstimulation of hair cells in the balance region. Sometimes the impulses that are fired from here go to the medulla oblongata instead of the cerebellum and trigger vomiting rather than affecting balance. (Remember, the medulla oblongata is in the brain stem and controls things like respiratory and digestive reflexes.) I've seen this in action at the playground—don't feed your child right before spinning her on the tire swing!

Thermoreceptors:

Receptors that sense heat are located all along the skin and in the blood vessels. The hypothalamus monitors the body's temperature, and if it is too low it will trigger responses to create and conserve heat: contract skeletal muscles to produce heat (shivering), contract muscles attached to hair follicles so that they stand on end (increase insulation), and divert blood from the extremities in severe conditions. If the temperature is too high, the body has responses to help cool it down as well (sweating, dilating blood vessels to let off heat that leaves a flushed color on the skin, panting, plus causing behavioral responses like taking off a sweater, getting in the shade, etc.).

Pain Receptors:

Pain receptors are called **nociceptors**, and there are several million of these everywhere except the brain. When the brain detects pain from these receptors, the neurons release "substance P," which triggers the release of endorphins and enkephalins (natural painkillers) to deal with the stimuli. Sometimes we can't really tell exactly where the pain is coming from when it is internal pain, because nerves from similar locations are entering the spinal cord at the same junction, and the spinal cord can't process which nerve the pain stimulus came from. One warning sign of a heart attack is pain in your arm. The pain is actually coming from your heart, but you process it as coming from your arm because nerves from both locations enter the spinal cord at the same spot. This is called "referred pain." When I had appendicitis, I felt the pain just above my stomach rather than where the appendix is located. That is another common referred pain that can signal that infection. (Too bad I didn't know that at the time!)

Photoreceptors:

Vision requires photoreceptors (**rods** and **cones**), which are located inside our eyes on the **retina**, a very thin layer of neural tissue lining the inner surface of the eyes. The central region of the retina at the back of the eye is called the **fovea** where most of the photoreceptors are located. The axons of the neurons from the retina converge at the optic nerve at the back of the eyeballs as well. Once the photoreceptors pick up the stimulus, the message will travel from the rods and cones → optic nerve → thalamus → occipital lobe, and then up for visual processing in the cortex of the brain (so that you understand what the vision means). The images we see get reversed in the eyes: upside down and backward, and then the brain corrects this.

The different photoreceptors are specialized for different wavelengths of light: **Cones** are for color vision, and there are three types: greens, reds, and blues (all colors are detected depending on which combinations of cones are stimulated). The hereditary disorder called **color-blindness**, which we will discuss further in the genetics chapter, means that a person cannot see certain colors because they do not make all three types of cones. (They are usually lacking the red and green cones.) **Rods** detect purple, black, and dim light (rods are important for night vision).

Signals from the right part of the eyes travel to the left hemisphere, and signals from the left part of the eyes go to the right hemisphere. The signals cross at the optic chiasmata in the brain (near the pituitary), and the message travels from there to the back of the brain in the occipital lobe. Images travel in through the lens of the eye, which is behind the iris (colored portion of the eye, with the pupil in the middle). Ciliary muscles on each side of the lens contract to adjust the shape of the lens so that the lens will direct the image onto the retina at the back of the eye.

Sometimes the muscles don't work properly and don't adjust the lens to focus the image directly onto the retina. The image can end up a little in front or behind the retina, which leads to near- or farsightedness. This can also happen if the lens is not shaped correctly or if it is too close to or far from the retina (**astigmatism**). Any of these irregularities can lead to vision problems, which can be corrected with glasses/contacts.

Vision problems increase with age as well, since the lenses add layers to themselves, which make them thicker and stiffer (harder for the ciliary muscles to shape the lenses to focus the image on the retina, which usually leads to farsightedness).

The parts of the eyes from the outside in:
- **Cornea**: The transparent layer over the entire front of the eye—protects the eyes, and helps to focus light.
- **Sclera**: The whites of the eyes, also help to protect the eyes; a **fibrous layer**.
- **Iris**: Pigmented portion of the eye (pigment is genetically determined by DNA). The pigment and fibrous muscle tissue form a unique pattern like a fingerprint.
- **Pupil**: The dark portion inside the iris is where light enters the eye. Circular muscles contract to shrink the pupil when we don't want to let in as much light, and radial muscles will contract to enlarge the pupil (important for night vision to let in more light).
- **Lens**: Located behind the pupil, focuses the light onto the photoreceptors.
- **Ciliary body**: On each side of the lens to shape the lens to focus images on the retina. It also secretes the aqueous humor.
- **Aqueous humor**: Fluid on both sides of the lens delivers nutrients to and takes wastes from the cornea and lens (since the choroid is not providing vascular functions to this area).

- **Vitreous humor**: Clear jellylike fluid inside the eyeball; 99% water with some sugars, salts, collagen derivative, and phagocytes (to remove cellular debris); maintains the shape of the eye and keeps the retina lying flat on the choroid.
- **Choroid**: Contains blood vessels to service the eye (located just to the outside of the retina); a **vascular layer**.
- **Retina**: The inner lining of the eye behind the vitreous humor, sensitive to light due to the photoreceptors that are located on the retina, mostly packed in the <u>fovea</u> at the center rear of the eye. The retina is the **nervous layer** of the eye, since this is where light is converted to nerve impulses because all the sensory receptors are located here. The impulses then travel to the brain via the **optic nerve** that is coming out of the back of the retina.

Disorders of the eyes

Color blindness—a sex-linked hereditary trait, which we will discuss more in Chapter 17 (Genetics). If a person is missing all three types of pigment proteins in cone cells, that person is totally color-blind and sees only in black and white and shades of gray. The majority of color-blind people are missing only the pigment proteins for the red and green cones, and therefore are unable to detect only those two colors.

Night blindness—occurs when there is a lack of vitamin A in the diet. Vitamin A is necessary to form retinal visual pigments prevalent in the rod photoreceptors (important for dim light).

Astigmatism—also inherited since it has to do with the shape of the eye, which affects vision; the cornea in one or both eyes has an irregular curvature and can't focus the light properly (Lasik surgery can reshape the cornea for nearsightedness, called myopia).

Conjunctivitis—infection due to bacteria, viruses, or allergies, which causes inflammation in the transparent membrane called the conjunctiva, which covers the sclera (white of eyes) and the lining of the lids. It is also called "pinkeye," since the sclera get very red during the inflammation (as blood vessels are more prominent as the body is trying to combat the infection; the body also produces extra mucus during infection for defense, which dries up to be crusty in the morning, making it hard to open the eyes). If a bacterium is the cause, it can be treated with antibiotics; viral conjunctivitis is harder to treat. Some STIs cause pinkeye as well.

Malignant melanoma—the most common form of eye cancer develops in the choroids (where the blood vessels nourish the eyes). Sometimes it is not noticeable until it has traveled to other parts of the body. Cancer can also occur at the retina, and treatment is usually removal of the eye.

Cataracts—proteins that make up the eye's lens sometimes lose shape/function as we age, and this causes clouding of the lens.

Macular degeneration—portions of the retina degenerate and scar tissue forms, creating blind spots.

Glaucoma—if too much aqueous humor builds up inside the eyeball, the blood vessels in the choroid can collapse from the pressure. If blood is not delivering oxygen and nutrients to the neurons of the retina, they die, and vision deteriorates.

Retinal detachment—if the retina tears for some reason (due to a concussion or illness), the vitreous humor can leak out of the eyeball, which causes the retina to become detached from the choroids. Neurons will not get the proper blood supply and vision damage or loss can occur.

Eye Twitch—can be from stress, caffeine, excessive alcohol, hormonal issues, poor nutrition, fatigue, or eyestrain. It is due to misfiring of neurons, which make muscles in the eyelid contract irregularly.

Keratoconus—a degenerative eye disorder in which the cornea has a conical shape rather than the usual curve and causes vision problems and sensitivity to light. The cause is still unclear but this condition may be caused by genetic factors when certain enzymes in the cornea are not functioning properly or possibly by environmental factors.

And just for fun …Why do we cry when we chop an onion? This is because onions release an enzyme that produces a gas that irritates our eyes by reacting with tears to produce sulfuric acid, which stings. The pain signals the brain to cause the release of more tears to flush out the acid. The bottom of the onion has the most of these enzymes, so chop from the top portion first. You can also reduce the effect by putting the onion in the fridge before you chop it, since the enzymes don't work as well in cold temperatures and won't release as much of the gas.

Chapter 15

Endocrine System

The endocrine system is unique in the fact that it is not a cluster of organs working together on a specific function. Instead it is composed of many different glands <u>and</u> organs scattered throughout the body working on many different functions that relate to all of the other organ systems. The endocrine glands and organs secrete chemical messengers called <u>hormones</u> directly into the blood (not through ducts like exocrine secretions). These hormones travel to their target cells to cause a response. (One hormone can have different effects depending on which target cell it travels to—and hormones have multiple target cells.) The endocrine system does not work as fast as the nervous system, but it is very similar in the way it can cause responses all throughout the body, affecting almost every cell! The endocrine system is regulated by the nervous system, so the two go hand in hand. Hormones are regulated by positive or negative feedback mechanisms, where the release of one hormone triggers or inhibits the release of other hormones, or hormones are triggered directly by nerve stimulation. Hormones are incredibly potent and have strong effects in minute quantities, so they are regulated in very narrow limits to maintain homeostasis; small imbalances can have large consequences.

Most of the hormones that we will go over in this chapter are part of the endocrine system, but I should also make the distinction that there are other chemical messengers that also cause a response in their target cells but are not part of the endocrine system. Instead of traveling in the blood to cause a response elsewhere, they act locally where they are made. So let's start by defining all three chemical messenger systems:

Three ways chemical messengers can trigger a response:

1. **Endocrine action**: Hormones are secreted directly into the circulatory system and travel to target cells in a different location from where they are made.
2. **Paracrine action**: Chemical compounds that act locally where they are produced (ex: prostaglandins and histamines). They become inactive a short distance from where they are secreted.

3. **Autocrine action**: Chemical messengers act on the same cell that produces it (the autocrine agent travels outside of the cell that produces it and attaches to a receptor on the same cell). An example is the protein cytokine interleukin-1 in monocytes (in the immune system).

Now that we have that distinction out of the way, we can look at the different types of hormones in the endocrine system and how they work.

Two Types of Hormones

1. **Steroid Hormone**: Hormones made from lipids (cholesterol derivatives). Because they are lipids, they are lipid-soluble, and since cell membranes are phospholipids, these hormones can pass through the cell membrane and nuclear membrane directly to the DNA of a cell to turn on and off sequences that code for proteins. The mRNA will make the new combination of proteins coded with the help of the hormone, and these new proteins alter the activity of the cell. There are multiple steps involved for these hormones to cause a response, so they are considered slow-acting hormones. Examples include estrogens, androgens, mineralocorticoids, and glucocorticoids.

2. **Non-Steroid Hormones**: Made from proteins or amino acids (sometimes called "peptide hormones"), which are water-soluble and are easily transported in blood or interstitial fluid. They <u>cannot</u> pass through the cell membrane of their target cell, so they attach to a receptor on the outside of the cell membrane and trigger a messenger (usually cyclic AMP) to go into the cell and trigger an enzyme (usually kinase) to cause a response in biochemical reactions. This process is faster than the steroid mode since it is going straight to the proteins that are already made and altering their functions (therefore altering the function of the target cell), rather than making that extra step of altering how the proteins are made. Examples include epinephrine, insulin, glucagon, thyroid hormones, etc. (All of the hormones in our body are proteins other than our sex hormones, prostaglandins, and corticoids.)

Hormone Major Functions

1. Organ function
2. Growth development
3. Reproduction and sexual characteristics
4. How body uses and stores energy (metabolism)
5. How sugar and salt balance in blood
6. Fight-or-flight sympathetic responses

Glands, Organs, and Hormones

(Starting from the top of the body and working down.)

In the brain:

- **Hypothalamus**: Produces eight different hormones. Two of these hormones are stored and released from the posterior pituitary lobe (ADH and Oxytocin). The other six are releasing and inhibiting hormones, which are released from neurons directly into the capillaries that connect with the pituitary capillaries to cause a quick response in the anterior pituitary gland.

o Hormones from the hypothalamus help to maintain homeostasis by regulating blood pressure, heart rate, water levels, and temperature. They also control circadian rhythms, emotional states, and reproductive functions. A very important protein called **kisspeptin** triggers the release of gonadotropin-releasing hormone (GnRH) from the hypothalamus, which then stimulates the follicle-stimulating hormone (FSH, in the anterior pituitary) to trigger the release of the sex hormones from the gonads (testes and ovaries). This protein starts the chain reaction that causes puberty and is coded for by the "Kiss 1" gene. The timing of when kisspeptin triggers GnRH depends on your DNA and biological clock, but it can also be affected by environmental factors such as diet.

o The six releasing and inhibiting hormones made in the hypothalamus that control the anterior pituitary hormones are
  – Corticotropic-releasing hormone (CRH)—affects ACTH in pituitary
  – Thyrotropin-releasing hormone (TRH) (also called prolactin-releasing hormone: PRH)—affects TSH
  – Gonadotropin-releasing hormone (GnRH)—affects FSH
  – Growth hormone-releasing hormone (GHRH)—affects GH
  – Prolactin-inhibiting hormone (PIH) (also known as Dopamine)—affects PRL
  – Growth hormone–inhibiting hormone (GHIH) (also known as Somatostatin: SS)—affects TSH and GH

- The **pituitary gland** is called the master gland since it makes many hormones and controls many glands. It is the size of a pea, and is situated right below the hypothalamus in the limbic system of the brain, linked by a slender stalk called the infundibulum.

- **Posterior Pituitary Lobe** stores two hormones that are made in the hypothalamus and secreted in the posterior pituitary lobe by neurons until they are needed. The release of these hormones is controlled by nerve stimulation.
  o Anti-Diuretic Hormone (ADH) (also known as vasopressin): Responsible for the reabsorption of water in collecting ducts of nephrons in kidneys. (This is secreted if the hypothalamus detects low water levels, therefore signaling that the body needs to conserve more water.)
  o Oxytocin: Important for uterine contractions and lactation (helps with childbirth by initiating labor and causing the milk "let-down" response during lactation). This hormone is also called the "bonding hormone" because it can be released during sex, hugging, and nursing to form bonds between individuals. It can also cause the release of endorphins, which are natural pain killers, in certain situations (like labor).

- **Anterior Pituitary Lobe**: Makes its own hormones. These first four hormones are messengers to stimulate other glands to release hormones, while the last three act directly on target tissue. Many of these anterior pituitary hormones are controlled by the hypothalamus-releasing and -inhibiting hormones (see above).
  o Thyroid-stimulating hormone (TSH): Stimulates development and activity of thyroid gland and activates the thyroid to produce the T3 and T4 hormones to regulate metabolic rate.

- o <u>Adrenocorticotropic hormone</u> (ACTH): Stimulates adrenal cortex glands to produce their hormones (corticoids); also affects fearlessness and aggression.
- o <u>Follicle-stimulating hormones</u> (FSH): Stimulates gonads—ovaries and testes—to grow and mature and to produce egg and sperm.
- o <u>Luteinizing hormone</u> (LH): Stimulates gonads to produce sex hormones (testosterone, estrogen).
- o <u>Growth hormone</u> (GH): Stimulates cell division and growth all over the body—muscles, bones, etc. It also promotes protein synthesis and influences height.
- o <u>Prolactin</u> (PRL): Stimulates milk production and lactation and promotes glandular tissue to increase during pregnancy.
- o <u>Human growth hormone</u> (HGH): Stimulates fat metabolism and growth.

- **Pineal gland**: Produces and releases <u>melatonin</u>; regulates sleep cycles governed by light (pineal cells are indirectly sensitive to light signals carried by the adjacent optic nerve).

- **Thyroid gland**: (front of throat) Secretes hormones to regulate metabolism and calcium levels; 95% of thyroid hormones are T4 and 5% are T3.
  - o <u>Thyroxin</u> (T4): Speeds up metabolism; responsible for cellular conversion of glucose to ATP by stimulating the production of proteins and increasing the amount of oxygen cells used to increase cellular respiration rates.
  - o <u>Triiodothyronine</u> (T3): Also helps with cellular respiration rates. Every cell in our body requires T3 and T4 for this purpose.
  - o <u>Calcitonin</u>: Lowers calcium levels in blood by putting it into the bone (uptake by osteoblasts), helps build bones, increases bone mass, and inhibits osteoclasts from breaking down bone.
  - o **Hypothyroidism** is when this gland secretes too little of T3 and T4, causing a slowing of the metabolism (tendency to gain weight), slow heart rate, low body temperature, fatigue, dry hair and skin, cognitive and memory problems, and muscular weakness.
  - o **Hyperthyroidism** (Graves' Disease) is too much T3 and T4, and causes increased heart rate, increased temperature, sweating, irritability, and weight loss.
  - o Women are 5–8 times more likely than men to have thyroid problems.
  - o Thyroid hormone production requires iodine from diet; if there is an iodine deficiency, it leads to goiter, an enlarged thyroid gland and neck, and reduced thyroxin and calcitonin production.

- **Parathyroid gland**: (back of neck) Releases <u>parathyroid hormone</u> (PTH), which raises blood calcium level by breaking down bone tissue, preventing loss of calcium from the kidneys, and stimulating calcium uptake from the digestive tract. PTH also regulates other electrolytes like phosphates in blood and is very important for muscle and nervous systems. High PTH levels can cause too much calcium in the blood, which leads to soft bones, fatigue, kidney stones, and personality changes. Low PTH levels cause problems with nerves and muscles such as spasms and tetany (continuous muscle contractions); twitches can result from low calcium levels.

- **Bronchioles**: Tubules in the lungs also secrete the paracrine chemical messengers <u>prostaglandins</u>: these help with muscle contractions of the tubules. (Prostaglandins are secreted in multiple areas and are important for defense, body temperature, controlling stomach acid, helping to move food through the digestive tract, and reproductive functions, which we will discuss below.)

- **Thymus gland**: Secretes <u>thymosin</u>, which stimulates the development of your T-cells (right above your heart).

- **Heart** has special cells in the atria, secreting <u>atrial natriuretic peptide</u> (ANP); this controls your blood pressure (one of a few) and is also a thirst suppressor.

- **Liver** secretes <u>leptin</u>, an appetite suppressor.

- **Stomach** secretes <u>ghrelin</u>, an appetite stimulator, and <u>gastrin</u> to make pepsin.

- **Adrenal glands** (above kidneys): Have two parts to them, the outer **cortex** is regulated by ACTH, released from the anterior pituitary (controlled by CRH from the hypothalamus) and produces and releases the corticoid hormones: glucocorticoid, involved in glucose metabolism, mineralocorticoid to maintain ion balance, and gonadocorticoids (androgens and estrogens), which are precursors to estrogen and testosterone. The **medulla** portion of the adrenal glands secretes epinephrine and nor-epinephrine and these are stimulated by sympathetic nerves during fight and flight (stress) responses.
  - o **Adrenal Cortex**: All of the hormones produced here are lipids and essential to life.
    - <u>Glucocorticoid = Cortisol</u> promotes glucose production, raises blood glucose levels, helps regulate blood pressure, and reduces inflammation. This hormone is released under stressful conditions to help the body react to the situation. Too much cortisol, however, suppresses the immune system, causes hyperglycemia, high blood pressure (which is dangerous for the vessels and heart), and increases acidity in the abdomen (which can result in ulcers). Stress reduction is important to keep this hormone in balance.
    - <u>Mineralocorticoid = Aldosterone</u> (in loop of Henle in kidneys) regulates salt concentration in nephrons by activating the pumps to remove salts from nephron tubules.
    - <u>Gonadocorticoid = Dehydroepiandrosterone (DHEA)</u>—the most abundant steroid hormone in humans. This acts on androgen receptors to produce testosterone and estrogens. It is also an important hormone to boost the immune system, decrease depression, promote clarity of thought and energy, and may help with skin complexion. High levels of DHEA have been linked to higher risk of hormone-sensitive cancers such as breast, prostate, and ovarian. Levels tend to start decreasing after age 30.
  - o **Adrenal Medulla**
    - <u>Epinephrine (aka adrenaline), and norepinephrine</u> (fight-and-flight responses when under stressful situations). Raises heart beat, slows digestion, dilates airways, etc.

- **Kidneys** release the hormone <u>erythropoietin</u> (EPO), which stimulates stem cells in bone marrow to make more red and white blood cells. It also releases the enzyme <u>renin</u>, which affects blood pressure/blood volume, and <u>calcitriol</u>, which increases the level of calcium in the blood by stimulating calcium and phosphate absorption (it is the hormonal version of vitamin D). It works with PTH and suppresses calcitonin from being released.

- **Pancreas**: Has both exocrine and endocrine glands (pancreatic islets with alpha, beta, and delta cells). Exocrine secretes digestive enzymes into the small intestines via ducts. Endocrine glands secrete hormones directly into the blood vessels:
  - Insulin lowers blood sugar by stimulating cells to take up glucose for cellular respiration to make ATP (energy). Insulin goes to work after every meal to clear the blood of glucose. (Insulin is produced and released from pancreas beta cells.) Diabetes results from insufficient insulin production or response and is the most serious chronic disease in the U.S. It affects 7% of the population and accounts for $92 billion in healthcare costs and causes ~225,000 deaths per year (sixth-leading cause of death).
    - **Type I Diabetes Mellitus** is an autoimmune disease, which attacks the beta islets cells and production of insulin ceases. From 5 to 10% of diabetes cases are Type I, versus 90–95% Type II.
    - **Type II Diabetes** occurs later in life when a person becomes insulin resistant or the beta cells stop functioning properly.
    - If not enough glucose is taken up by cells, that means there is more glucose in the blood, which means the kidneys will lose too much water when excreting the excess sugar. This leads to dehydration, low blood pressure, weight loss, and more acid in the blood, since the cells are constantly breaking down fat to make up for not having enough glucose. Acidosis can damage brain function as well. Diabetes is a very serious disease, which roughly 21 million Americans have. Of those, 6.2 million people are unaware that they even have it.
    - **Gestational Diabetes** occurs in pregnant women who often have no history of diabetes I or II but develop high blood sugar during the end of pregnancy. This happens if the woman does not secrete enough insulin required to regulate her blood sugar. It is easily treated, but if not treated, the woman will have a higher risk of diabetes after pregnancy and the child will also have a higher risk of obesity and diabetes later in life.
  - Glucagon is secreted from the pancreas when glucose levels are low. Glucagon raises the level of sugar in your blood by converting glycogen to glucose in the liver. It also breaks down glycogen in muscles and stimulates the production of glucose from amino acids. It increases blood sugar between meals, providing energy to active muscles and the brain. (Glucagon is produced in and secreted from alpha cells.)
  - Somastatin: Regulates insulin and glycogen secretion.

- **Small intestine**:
  - PYY: An appetite suppressor.
  - Cholecytokinins: Stimulates the contraction of the gallbladder to release bile.
  - Secretin: Stimulates the pancreas to release bicarbonate-rich juices to neutralize the chyme as it enters the small intestines from the stomach.

- **Ovaries**: Produce eggs and sex hormones.
  - Estrogen: Important for egg maturation and release, prepares uterus for pregnancy, regulates breast cell division and development, helps regulate menstrual cycle, helps to maintain the heart and healthy bones.

- o <u>Progesterone</u>: Maintains endometrial lining of uterus during pregnancy, helps to regulate menstrual cycle and breast development.
- o <u>Testosterone</u>: Important for libido/desire for sex.

- **Uterus**: Produces <u>prostaglandins</u> for uterine contractions (helpful also to move sperm up for fertilization). Prostaglandins here are also the cause of menstrual cramps.

- **Placenta (during pregnancy)** is a temporary endocrine gland that secretes <u>chorionic gonadotropins</u>, which help to maintain the **corpus luteum** (a tissue derived from the follicle that secretes progesterone and estrogens to help maintain the endometrial lining and trigger mucus production from the cervix).

- **Testes**: Produce …
  - o <u>Testosterone</u>: Stimulates sexual characteristics, develops genitals, and triggers sperm formation. Testosterone affects sex drive, chest and facial hair, pubic hair, muscle mass (doubles during puberty), and cartilage. The cartilage in the vocal folds widen, lowering boys' voices during puberty, and the Adam's apple develops more. The testes also responds to hormones that are usually associated with the female reproductive system, but these hormones have functions in the male reproductive system as well:
  - o <u>Estrogen + Progesterone</u>: Necessary for proper sperm development.
  - o <u>Prolactin</u> (in pituitary): Important for testosterone synthesis.
- **Seminal vesicles**: Secrete <u>prostaglandins</u> into the sperm for muscle contractions.

I've already mentioned some of the problems that occur when hormones are not produced or regulated properly (hyper/hypothyroidism and diabetes), but there are other hormonal changes that occur due to the aging process. Women go through menopause due to the decrease in estrogen over time. They stop menstruating, and the hormonal change can also cause hot flashes, chills, sleeplessness, irritability, and bone loss. Men's testosterone levels also decline as they age, but the effects are less dramatic. Sperm count and sex drive may go down, but men can still father a child well into their 70s. For both sexes, decrease in growth hormone will affect bone and can cause more fat production; aldosterone usually decreases by 30% by age 80, which can cause dehydration and urinary problems. The parathyroid hormone levels increase with age and lead to bone loss and osteoporosis.

Chapter 16

Cell Division and Reproductive System

The reproductive system is the only system not involved in homeostasis in the body—the only function is to continue our species. The reproductive system consists of reproductive organs such as the gonads (testes and ovaries), accessory glands, and ducts. These function together to produce gametes (sperm and eggs); sustain, transport, and unite these cells to produce a new individual; produce and release sex hormones to help with the reproductive process; and then support and nourish the developing offspring. Before we start the reproductive system, I want to cover some vocabulary that you will need to know to understand reproduction. In order to understand how we make our gametes (egg and sperm) in our gonads, we need to first discuss **cell division**, the process of replicating our DNA and creating new cells. Roughly 25 million cells divide per second in an adult human; this enables us to grow and develop as well as repair and replace cells, all the time. Cell division is very controlled and once a cell divides 50–60 times, it will stop dividing. (Uncontrolled cell growth is called cancer, which we will go over in Chapter 18.) There are two types of cell division. **Mitosis** produces cells identical to the parent cell. We are constantly creating cells by mitosis to replace our skin cells, our cells lining the stomach, blood cells, etc. **Meiosis** is a very special cell division, which only occurs in the gonads to produce our gametes, and results in cells with <u>half</u> the number of chromosomes from the parent cell. We will briefly go over the two types of cell division here.

All of our body cells other than our gametes are called **somatic** cells. Somatic cells undergo the cell division called **mitosis** to produce diploid cells identical to the parent cell. **Diploid** means that chromosomes are in pairs. Our somatic cells have 23 <u>pairs</u> of homologous chromosomes, which means 46 chromosomes total in every somatic cell. (Homologous means that the chromosomes in each pair are the same size and have the same banding pattern where genes occur. The DNA sequences of the homologous pairs will be slightly different because the chromosomes may have different alleles (alternate forms of genes). For example, one allele may code for freckles and the other allele on the homologous chromosome may code for no freckles, but they are both coding for that particular trait in that particular segment of the DNA. (We will discuss this more when we get to Genetics in the next chapter.) During mitosis, the parent cell replicates its DNA in the nucleus and then divides so that the two new cells have the same DNA that the parent cell started with. The phases of mitosis are Interphase, Prophase, Metaphase, Anaphase, Telophase, and Cytokinesis. The basic process is as follows:

Mitosis

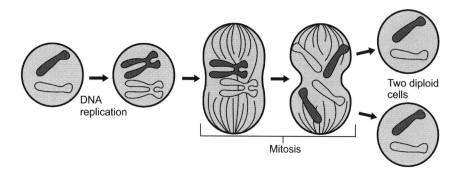

- For 90% of a cell's cycle, it is NOT dividing, but is growing and preparing for division. This phase of the cell cycle is called **Interphase** and is divided into three portions (two "gap" phases [G] and one "synthesis" phase [S]):
- G_1 phase—cell grows.
- S phase—chromosomes duplicate (although at this point they are not discrete condensed "chromosomes" yet, the DNA is in an uncoiled formation called "**chromatin**").
- G_2 phase—cell grows more and prepares for division. After Interphase the cell starts the Mitotic Cycle, which includes the following phases:
- **Prophase**—The chromatin becomes tightly coiled and condenses into discrete chromosomes, which you can see with a microscope. The replicated chromosomes are called "**sister chromatids**" and they are joined at the **centromere** (a region at the center of the sister chromatids). The mitotic spindle begins to form between two **centrosomes** (consists of centrioles and asters, which make microtubule spindle). The centrosomes start to move to opposite sides of the cell, with the spindle forming across the cell.
- **Prometaphase**—the nuclear envelope fragments so that the sister chromatids can now interact with the spindle that has formed across the cell. The chromatids attach to the spindle at their centromeres.
- **Metaphase**—longest phase of mitosis; the sister chromatids line up in the middle of the cell at an imaginary equator called the "**metaphase plate**."
- **Anaphase**—paired centromeres of each chromosome separate, meaning that they are no longer "sister chromatids" anymore, but each half of the sister pair is now a full-fledged chromosome. These new chromosomes start to move toward opposite poles with centromeres leading the way (the portion that used to attach the two portions together to make them "sisters" is now on both of the new chromosomes). The new chromosomes are still attached to the spindle.
- **Telophase**—the cell elongates even more, and the two daughter nuclei begin to form at the two poles where chromosomes have accumulated. A nuclear envelope forms from fragments from the parent

cell's nuclear envelope and the nucleoli reappear. Chromatin becomes less tightly coiled, regains the form it had before mitosis occurred.

- **Cytokinesis**—is the splitting of the cell into two new cells, each identical to the other in this case, with the same number of chromosomes as the original parent cell. In plant cells, a cell plate forms across the dividing cell, and in animal cells, a "cleavage furrow" occurs, pinching the cells into two. That is really the quick and simplified version of mitosis. Now for even more of an outline for meiosis:

Our **gametes**, on the other hand, are **haploid** cells, which means that chromosomes are not in pairs. (There are only 23 chromosomes in eggs and sperm. When they unite at fertilization, the resulting zygote restores the diploid condition. So the pairs of chromosomes are due to the fact that you get one whole set of 23 chromosomes in the egg from your mom and the other set of 23 chromosomes from your dad carried in the sperm and then released into the egg.) As I've mentioned, mitosis produces diploid cells that are identical to the parent cell.

Meiosis

Meiosis is another type of cell division that only occurs in gonads to make haploid cells (no homologous pairs, just 23 chromosomes). This is very important for reproduction. Once the haploid cells unite at fertilization, they restore the diploid condition. There are special diploid progenitor cells in the gonads that go through meiosis to produce gametes—sperm and egg. (Sperm are made in the testes, and eggs are made in the ovaries.) Meiosis is different from mitosis. Instead of the basic phases listed above, meiosis goes through all of these phases (other than cytokinesis) twice, so there is meiosis phase I and meiosis phase II in order to get down to 23 chromosomes. In the process, meiosis generates four cells rather than the two that mitosis generates, and the four daughter cells are not identical to the parent or to themselves; another very important difference, since this is what helps to create genetic diversity among individuals.

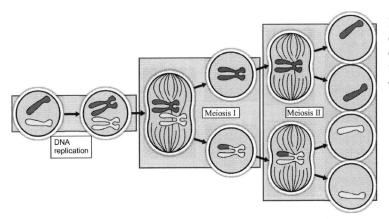

Here is a very simplified depiction of a diploid cell with only 1 pair of chromosomes going through meiosis to form 4 haploid cells with one chromosome each.

- **Meiosis**—a form of nuclear division that produces daughter cells having half the number of chromosomes found in parent cells; i.e., humans—46 chromosomes in somatic cells (23 pairs) means that through meiosis the gametes (egg and sperm) will get 23 chromosomes each.

- o **Parent cells are diploid** (nucleus contains pairs of chromosomes). They are special "progenitor cells" that create the gametes. These parent cells are called **Spermatogonium**—yields 4 sperm cells;
- o **Oogonium**—yields an oocyte (which becomes the egg) and three polar bodies (which are small and disintegrate eventually). The size of the large egg versus the small polar bodies is due to an uneven distribution of cytoplasm during **cytokinesis**.
- **Gametes are haploid** (nucleus only has one chromosome from each pair).
  - o Haploid gametes fuse to produce diploid zygote;
  - o Fertilized egg = **zygote**;
  - o At fertilization, zygote contains one set of chromosomes from mom and one set from dad;
  - o New zygote has homologous chromosomes.
  - o Meiosis produces gametes that contain a mixture of maternal and paternal chromosomes—this leads to variation.
  - o Other causes of variation are independent assortment and **crossing over (recombination)**.

Looking at the stages of meiosis to create sperm and egg:
- **Interphase I**—like mitosis, there is an interphase with an S and two G phases. Each set of chromosomes replicates during interphase I.
  - o Two identical chromatids attached at centromeres and centriole pairs replicate.
- **Prophase I**—The biggest difference between meiosis and mitosis occurs here. Chromosomes condense and attach at their ends to the nuclear envelope;
  - o Synapsis occurs—homologous chromosomes (each made of two chromatids) come together as pairs called **tetrad**;
  - o This is where crossing over can occur (this is pretty rare);
  - o Segments of adjacent non-sister chromatids may exchange DNA segments by breaking and reattaching to the other chromatid;
  - o Site of crossing over (where non-sisters are linked) called **chiasmata**;
  - o Chromosomes then detach from the nuclear envelope, centrioles move to opposite poles, and spindle forms;
  - o Chromosomes begin to head toward metaphase plate;
- **Metaphase I**—tetrads align at metaphase plate;
- **Anaphase I**—sister chromatids remain attached at centromeres and move as unit toward poles; homologous chromosomes move toward opposite poles;
- **Telophase I**—cleavage furrow or cell plate breaks into two cells;
- **Interphase II**—**no replication**—this is important, because when the cells divide a second time, they will get down to half the number of chromosomes as the parent cell due to the fact that the cells split twice but replicated only once.
- **Prophase II**—no tetrads now; they are already condensed, they just get in position for the next phase.
- **Metaphase II**—sister chromatids align on the metaphase plate;

- **Anaphase II**—sister chromatids separate and move toward opposite poles (looks like two hands with fingers toward each other and wrists away from one another).
- **Telophase II**—Similar to mitosis above, and then finally …
- **Cytokinesis** will occur in both of the two cells that are going through the second meiotic phase and result in four new, genetically different cells.
    o **Homologous** chromosomes = chromosome pairs of the same length, centromere position and staining pattern that possess genes for the same characters at corresponding loci;
    o One homologous chromosome is from mom and one is from dad;
    o Diploid number 2n = # of pairs;
    o Independent assortment—the orientation of the homologous pairs relative to the poles of the cell are random (during metaphase/anaphase I);
    o There are two alternating possibilities for each pair. Each pair separates independently from all other pairs, so you end up with genetic variation (random assortment of maternal/paternal chromosomes);
    o There are 8 million possible chromosome assortments (2^{23}, two possibilities for each pair);
    o The progenitor cells in the testes that are the parent cells for sperm are called **spermatogonia**.
    o Primary meiosis makes two sperm.
    o Secondary meiosis makes four sperm.
    o Mitochondria makes ATP (fuel packets).

For every one **male progenitor cell**, you make **four sperm**.
- **Spermatogenesis** takes about 74 days to complete in the seminiferous tubules of the testes.
    o The female progenitor cell, **oogonium**, goes through two stages of meiosis to produce one large egg and three small polar bodies due to uneven cytokinesis (when the cells divide, more cytoplasm goes into the daughter cell that becomes the egg). The egg is not actually considered an egg until fertilization, so looking at the phases of meiosis for the oogonium, it goes like this:
    o Primary meiosis makes one primary oocyte and one polar body.
    o Secondary meiosis makes two polar bodies and the primary oocyte becomes a secondary oocyte.
    o The result is three polar bodies (which disintegrate) and one secondary oocyte, which becomes an egg upon fertilization (during ovulation, the secondary oocyte stops at metaphase II and then continues to complete meiosis II if fertilization occurs.)

At fertilization, the sperm (which has 23 chromosomes) plus the egg (with 23 chromosomes) become a zygote with 46 chromosomes; the diploid condition is restored since they are now back to being homologous pairs of chromosomes (23 pairs = 46 chromosomes).

Now let's look at how these haploid cells unite with one another to form a new individual. It's an incredible system!

Male Reproductive System:

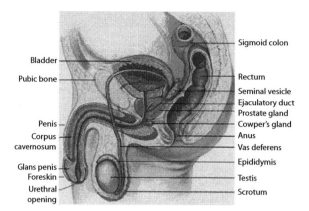

Testes: Produce sperm and **testosterone**. (Testosterone is the direct trigger for sperm formation, but other hormones like estrogen and LH also help with sperm formation, and prolactin helps with testosterone formation.) Males start developing sperm ~14 years of age until death, and in a lifetime produce about 400 billion sperm (around 100 million/day). Each sperm takes about two months to mature, but they won't be able to fertilize an egg until they have actually gone through the woman's reproductive tract (or have been manipulated if used for IVF). Sperm production starts in the seminiferous tubules in the testes from male germ cells called spermatogonia, which start to go through meiosis (cellular division that halves the number of chromosomes of the germ cells to produce sperm). Immature sperm have all the parts but are unable to move or fertilize an egg yet. Since sperm are haploid cells, they are considered invaders and are often attacked by immune cells. There are "**Nurse cells**" in the tubules to protect, feed, and move sperm to the traveling canal where they will go from the seminiferous tubules to the epididymus to mature. Each nurse cell cares for about 150 sperms! They also engulf degenerated sperm and produce some estrogen.

Sperm parts:
- **Acrosome**: Caplike structure, filled with digestive enzymes.
- DNA of sperm (23 chromosomes) in the head of sperm.
- Lots of **Mitochondria** and fuel (food packets) in middle portion of sperm.
- **Flagella tail** (made of microtubules) to help sperm swim up reproductive tract of the woman.

Epididymus: Sperm from seminiferous tubules are sent here to mature, and become mobile. Sperm are stored here until they are ejaculated. If ejaculation does not occur they will just disintegrate. They can stay in the epididymus for several weeks. If sperm is ejaculated (first going through the **vas deferens**), it picks up other molecules along the way to make the **semen**. Approximately 150–350 million sperm are ejaculated at once; 20% of these will be deformed with extra tails or severed heads, or attacked by bacteria. Deformities can be due to smoking, alcohol or drug abuse, chemical pollutants, radiation, poor nutrition, or high temperatures. (The temperature in the scrotum should be approximately 4 degrees lower than the male's body temperature, so hot tubs and tight clothing like jeans or tight-fitting underwear can have an adverse affect on sperm production and performance.) If sperm are not ejaculated within a few weeks, they just disintegrate and get reabsorbed.

Semen: Composed of sperm, fructose, prostaglandins, mucus, and buffers. Sperm are viable usually for only 24 to 48 hours after ejaculation, but some sperm can last for up to 5 days in the female reproductive tract! (One of the reasons the "rhythm method" doesn't work!) Of the millions of sperm that enter the female during ejaculation, only about fifty will actually make it to the egg, since there are so many obstacles along the way. Males need to release semen periodically in order to stay healthy. Many studies show that ejaculating at least twice a week when in your 20s can reduce the instances of prostate gland disorders, especially **prostatitis**, which is inflammation of the prostate.

Seminal Vesicles: Secrete fructose (for energy) and prostaglandins (triggers muscle contraction, which help with ejaculation as well as with sperm getting up female reproductive tract, as the prostaglandins make her muscles contract too).

Bulbourethral gland: (also called the Cowper's gland) The size of a pea and secretes mucus, which aids swimming sperm and lubricates and buffers the semen.

Prostate: Secretes buffers (the milky substance that makes up the bulk of the semen. This helps the sperm in the acidic environment of the vagina (pH = 3.5–4) so that some sperm will make it through the female reproductive tract (sperm need higher than 6 pH).

Penis: Three columns of erectile tissue that surround the **urethra**. This becomes engorged with blood during sexual arousal to cause an erection. (Remember from the nervous system chapter that this is due to the neurons in the brain releasing Nitric Oxide [NO], which triggers the blood vessels in the penis to dilate. Blood rushes into the vessels and causes the erection.) There is a bulbocavernosus muscle surrounding one of the columns of erectile tissue that contracts during ejaculation to expel the semen and at the same time stop the flow of urine. (A little urine may be mixed with the semen if there was any residual urine in the urethra at the time of ejaculation; usually not much, however.) There are other muscles in the penis as well to help with stabilization and ejaculation.

Female Reproductive System:

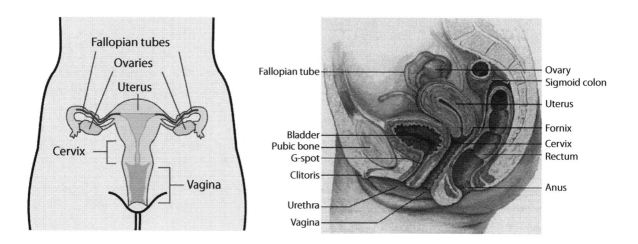

Ovaries: The primary reproductive organs in the woman. They are the size and shape of almonds and are located on each side of the woman's pelvic cavity above the uterus. They produce and secrete estrogen, progesterone, and contain the progenitor cells oogonium, which will go through meiosis I to produce oocytes (to become eggs). **Follicles** in the ovaries contain the developing oocytes, which are all made in a female when she is in utero (made when the female is a fetus). The oocytes are surrounded by nutritive cells to help feed and protect them. A female baby will have about 1 million oocytes in each ovary at the time of birth, but most disintegrate during childhood and there will be only ~300,000 follicles left once she reaches puberty.

ONLY 300–500 follicles fully mature in a lifetime during **ovulation** (when an oocyte is released from the follicle from the ovary). One follicle is released from one ovary every month from puberty to menopause. The oocyte does not mature to become an egg until after ovulation (then the primary oocyte will start the second phase of meiosis—it doesn't actually complete meiosis II unless fertilization occurs, however, so I will continue to call it an oocyte). If the oocyte is not fertilized, it is expelled (disintegrates) during menstruation, along with the endometrial lining of the uterus.

Fallopian tubes (Oviducts) are the critical 7–14cm-long passageway between the ovaries and the uterus where fertilization can occur. An oocyte is released into the space between an ovary and a fallopian tube (on one side or the other) and the fimbriae (fingers) of the fallopian tube sweep the **ovum** (oocyte and nutritive layers) into the fallopian tube. The ovaries are attached to the uterus by ligaments and the fallopian tubes reach up toward the ovaries from the uterus but are not connected to the ovaries. The ovum is the size of a grain of sand, and the passageway in the fallopian tubes is the width of two human hairs! It also has cilia to help move the ovum along. The journey through the fallopian tube takes about 7 days but the oocyte must be fertilized within 12–24 hours after leaving the ovaries in order for a zygote to form.

Uterus: Where the zygote develops into an embryo and then into the fetus. The fallopian tubes are attached to the upper portion of the uterus. Fertilization occurs in the fallopian tubes and the zygote travels down to the uterus and becomes embedded in the endometrial lining of the uterus. The uterus is usually the size and shape of a pear but it can expand tremendously to hold the developing fetus(es). Some of the mother's organs will be rearranged during pregnancy in order for this expansion to occur, some are pushed up into the chest cavity more, and the stomach is rotated 45 degrees to make room for the developing fetus.

Endometrial Lining: Where the zygote will implant in the uterus. It gives rise to the placenta, which nourishes the developing fetus and is attached by the umbilical cord.

Cervix: At base of uterus, connecting the uterus to the vagina. The cervix secretes thin mucus during ovulation to help with sperm delivery, and once implantation has occurred (or during menstruation), the mucus is thicker to help keep bacteria out. The cervix also helps to keep the fetus inside the uterus. During labor the cervix dilates so that the baby can move through the birth canal (vagina) to get out.

Vagina: A muscular, elastic chamber for penile insertion and is also the birth canal and passageway for the menstrual flow.

Bartholin's Gland: Beside the vaginal opening—produces mucus that helps with vaginal lubrication during sexual arousal. The main area for secreting mucus, however, is the cervix.

Vulva: Minor and major labia are external female genitals.

Clitoris: Eight thousand nerve endings are here, and the function is pleasure (it is the only organ of the body whose sole purpose is for pleasure). The tissue has the same embryological origin as penile tissue (it makes sense that if a female has an orgasm it is mostly from clitoral stimulation, just like male orgasm is from penile stimulation). During embryonic development the tissue will either become a clitoris or a penis depending on what sex hormones come into play (around 7 weeks). The hormones acting on our bodies in this regard are determined by which sex chromosomes we have: XX (female) or XY (male). Similar tissue also diverges to form scrotum versus labia, and both sexes get breasts and nipples despite the fact that females can produce milk from their breasts later in life and males cannot. (This also has to do with hormones).

Breasts are very important for nourishing the developing offspring after birth and also function in sexual attraction and pleasure. There are 4 hormones necessary for the proper production and ejection of milk: estrogen stimulates the development of adipose and glandular tissue in the breasts (milk glands are modified sweat glands), prolactin stimulates milk production within these glands, progesterone stimulates the development of the mammary duct system, and oxytocin causes the milk to be released from the ducts out of the nipples. Breast milk is the perfect combination of carbohydrates, fats, and proteins for the developing baby, and also contains the mother's antibodies to help fight infections since the baby does not have its own antibodies yet.

Let's look at how **fertilization** occurs (how the sperm nucleus and egg nucleus get together to make a new individual), including all of the hormones that help in the reproductive process:

LH (Luteinizing hormone) and FSH (Follicle-Stimulating Hormone) from the anterior pituitary gland stimulates the follicle to develop, and the growing follicle releases **estrogen**, which causes the cervix to secrete the thin cervical mucus that will act like channels to aid the swimming sperm. Estrogen also helps to ready the endometrium for pregnancy.

Another LH surge occurs right before ovulation and causes the follicle to release the oocyte with nutritive cells and zona pellucida (protective coat around the oocyte, which only one sperm can penetrate) into the oviduct. LH also triggers the remaining follicle tissue to become a mass of cells called corpus luteum, which secretes two hormones, a form of estrogen (estrodiol) and progesterone. The progesterone helps to maintain the uterus during pregnancy and the estrodiol triggers the cervix to secrete more mucus, which is sticky and thicker now in order to prevent bacteria from entering. (When the corpus luteum is in place, the hypothalamus signals for the decrease in FSH, which prevents other follicles from developing.)

IF… there is no implantation of a fertilized egg, the corpus luteum, oocyte, and endometrial layer will all be expelled after ~12 days during menstruation, also called a "period."

During ovulation the oocyte goes into an oviduct (either one), and needs to be penetrated by sperm within 12–24 hours of leaving the ovary in order for fertilization to occur.

Remember that sperm can last up to 5 days on the journey to find the egg but they usually get there 10–48 hours after ejaculation. It is a treacherous journey and most sperm will not make it to the oviducts. (This is one of the reasons men create so many sperm, to ensure that some will make it to the egg!) The sperm need to get past the very acidic vaginal area, where many die, and then the survivors will start swimming up to the uterus; they are either helped by the thin mucus channels (around the time of ovulation) or they are hindered by the thicker mucus secreted from the cervix at other times. The surviving sperm split up into the two oviduct channels and if there is an egg in one of the oviducts, the sperm that make it all the way will try to penetrate the zona pellucida. Out of the 150–330 million sperm ejaculated, only a few hundred will make it into the oviducts and then only about 50 will make it to the egg.

Each sperm has 23 chromosomes. The 23rd chromosome of both the egg and the sperm are the sex chromosomes. Eggs have only X chromosomes, since women are XX. The 23rd chromosome of the sperm can be either an X or a Y, since males are XY. Therefore it is the sperm that will determine what sex the baby will be, depending on which sperm fertilizes the egg. If the 23rd chromosome is a Y chromosome, the baby will be a male (XY). If the 23rd chromosome on the sperm is an X, the baby will be a female (XX). Some interesting generalizations about sperm (I'm not sure how scientific these data actually are): Sperm with Y chromosomes tend to swim faster but also die faster, and sperm with X chromosomes are usually slower but can remain in the females' reproductive tract longer. If this is indeed true, that would mean that women who have sex right around ovulation are more likely to have a boy since the Y sperm would make it up to the egg the quickest. Women who have sex a couple of days before ovulation might be more likely to have a girl since there would be more sperm with X chromosomes still alive searching for an egg.

If sperm make it to the oviduct (which is two hairs wide), chemical signals will guide them toward the oocyte. Many will encounter the oocyte at once. The oocyte is covered by a layer called zona pellucida, and the acrosome of the sperm will start releasing digestive enzyme to try to penetrate this layer. (The acrosome cap has been gradually breaking down during the sperm's trip up the female's reproductive tract, and will be ready to break apart by the time it encounters the oocyte.) The oocyte will complete meiosis to become an egg once a sperm penetrates the zona pellucida, and the egg triggers the zona pellucida to close so that only one sperm can enter. The sperm DNA from the head combines with egg DNA; this is **fertilization**, which makes the **zygote**, the first cell made upon fertilization.

The zygote travels down the oviduct to the uterus and starts dividing by mitosis from 2 to 4 to 8 to 16 cells, etc. (in a process called cleavage), until it is a ball of cells called a **morula**. The cells continue to divide and start to move toward the outside of the sphere. Once it has 70-100 cells toward the outside and a fluid filled cavity on the inside it is called a **blastocyst**. This is the mammalian term for a blastula. (All of the division is not increasing the size of the zygote yet; it is just increasing the number of cells the zygote has.)

The blastocyst has an outside layer of cells called the trophoblast that will develop into the placenta and amnion. The embryoblast mass, to the inside of the trophoblast, will develop into the embryo. The fluid filled cavity of the blastocyst is called the blastocoele. The trophoblast implants into the endometrial lining of the uterus within approximately ten days. The endometrium and trophoblast layers enclose around the embryo and develop into the amnion, which provides cushioning for the developing fetus; and the chorion, that forms the placenta, to allow for oxygen and nutrient exchange.

Gastrulation occurs after the blastula phase. This is when the blastocyst becomes a three layered **gastrula** that starts to form a gut. The three germ layers of the gastrula are the endoderm, mesoderm and ectoderm. These germ layers start to differentiate during the next phase, called organogenesis, to give rise to body tissues and organs.

Endoderm: Becomes the epithelial lining inside many systems such as the pulmonary, hepatic, urinary and digestive systems.
Mesoderm: Becomes the bulk of our tissue, developing into the different forms of muscles and connective tissues.
Ectoderm: Differentiates into the skin and the nervous system (brain, spinal cord, and nerves).

The corpus luteum is in place for a while to maintain the pregnancy hormones. It disintegrates once the placenta forms and starts producing estrogen and progesterone. The placenta is formed from the endometrial lining and the trophoblast after the embryo attaches and starts to develop. (The placenta has two components, the maternal placenta is from the endometrial lining of the mother and the fetal placenta is from the trophoblast of the blastocyst.) The umbilical cord is attached to the embryo/fetus, which is attached to the placenta. The mother's blood never enters the embryo; the nutrients and gases just exchange from the blood through the umbilical cord into the embryo/fetus.

Development in utero:

At 2 weeks: dividing cells become an embryo, which secretes the hormone Human chorionic gonadotropin (hCG) that can be detected in mom's urine (with a pregnancy test). (The primary function of hCG, secreted first by the placenta and then by the embryo, is to maintain the corpus luteum for a while so that it will continue to produce progesterone to maintain the pregnancy.) (If you have heard of hCG injections in association with losing weight, this is another "fad diet" ploy that should be avoided. It has no scientific benefit for weight loss.)

4–5 weeks: arm buds, nose, and eyes develop

6 weeks: embryo has legs

7 weeks: genitals to start to develop, as well as hands and fingers

9 weeks: now considered the **fetus** and about 2 inches long (has post-anal tail, a chordate trait). (Referred to as a fetus until birth when it is called a baby or infant.)

12 weeks: 3 inches long

14 weeks: can suck thumb

16 weeks: actively turn in uterus and senses are developing

18 weeks: 5.5 inches long; mouth, lips, and nasal plugs develop

24 weeks: can have viable birth, but many problems are associated with premature births (undeveloped lungs, poor vision, hearing loss, mental problems, etc.). Generally, the earlier the baby is born, the more prone it is to have health problems in the future. Some reasons for premature birth are uterine or cervical problems, no prenatal care, smoking, alcohol or drug use, poor nutrition, previous pre-term delivery, pregnancy with more than one fetus (twins, triplets, etc.), stress, domestic violence, and health conditions of the mother such as infections, obesity, high blood pressure, diabetes, or birth defects in the baby.

32 weeks: fully developed fetus, but delivery at this stage is still called premature birth and can still have associated risks

37–42 weeks: full term (everything has developed and the fetus is a good size for living outside the mother)

Beyond 42 weeks: induce labor, since larger babies are more dangerous for mother's and for baby's health

Twins: <u>Fraternal twins</u> (dizygotic) result from two oocytes being released from ovary/ovaries and two sperm fertilizing the two eggs and both eggs implanting in the uterus. These twins are just as genetically different from one another as regular siblings; they just happen to share a birthday and develop in the uterus together. Occurrence can be genetically linked to the mother (if there are twins on her side of the family, she might have twins too), but not to the father.

<u>Identical twins</u> (monozygotic) result from one oocyte being fertilized by one sperm and then the developing zygote (at the blastocyst stage) splits into two and develops into two embryos. The embryos have the exact same DNA, so they are identical (in every way except for their fingerprints and iris patterns). This is not genetically influenced; it is just a random occurrence.

We already discussed some **sexually transmitted infections (STIs)**, such as chlamydia, gonorrhea, syphilis, HPV, herpes, and AIDs in the Immune System Chapter, so let's go straight to …

Birth Control Methods

Methods to prevent pregnancy (and STIs in some cases)

The most reliable method, of course, is abstinence; but if that is not going to happen, then it's best to be prepared with a reliable birth control method so that if you are not ready to have a baby, you won't, and so that you and your partner are protected from sexually transmitted diseases and infections.

Surgical Sterilization: After abstinence, this is the next-best method for deterring pregnancy (but does nothing to prevent STIs). The failure rate for surgical sterilization preventing pregnancy is generally less than 1%.
- Vasectomy: The vas deferens are both snipped so that sperm cannot be ejaculated with the semen. The man can still ejaculate the rest of the semen (sperm only makes up roughly 1% of semen), but the woman will not become impregnated.
- Tubal Ligation: This is when the female's oviduct tubes are cut/tied. The oocyte released from the ovaries, therefore, can never be fertilized since the sperm can't reach the oocyte. This does not affect menstruation, because hormones are still being produced to signal the endometrial lining to build up and be shed each month, despite the fact that there is no oocyte traveling to the uterus.
- Hysterectomy: When the uterus is removed.
- Essure is when a coil is inserted inside the fallopian tubes to block them off.

Hormonal Contraception: Pills, patches, and vaginal rings with different combination of hormones to prevent pregnancy. They don't, however, guard against STIs. The failure rate for preventing pregnancy for these types of contraceptives is anywhere from 1–8%. The pill is the most commonly used contraception by women in the U.S.
- Estrogen and progesterone combo: Inhibit FSH and LH so that the follicle won't be released from the ovary and eggs don't fully develop.
- Progesterone only: Can prevent pregnancy by preventing ovulation, causing extra-thick cervical mucus (harder for sperm to swim through), and by inhibiting the proper preparation of the endometrial lining so that if the other two methods didn't work and the egg is fertilized, the zygote will not be able to implant properly.
- Side effects: Vary depending on the person, but some common side effects are increased risk of blood clots, high blood pressure (especially in smokers), weight gain, acne, and gallbladder problems. High estrogen levels can increase the risk of breast cancer. One good side effect is that hormonal contraception often decreases the flow of menstruation; certain kinds can decrease the number of times you menstruate within a year.

Intrauterine Device (IUD): Prevents fertilization or implantation or transport of blastocyst, does not protect against STIs. Some have progesterone and have side effects similar to some of the pills. Some increase menstrual flow and some decrease it. IUD failure rate is less than 1% and it can be left in for up to 5 to 10 years depending on which type you get.

Barrier Methods: All of the barrier methods are designed to stop the sperm from reaching the egg after ejaculation. Some, but not all, DO help to prevent STIs, so it's good to know which ones are the most effective at that as well.

- Diaphragm protects only from cervical genital warts (which can cause cervical cancer). Failure rate for preventing pregnancy is 15–20%.
- Cervical Cap: Protects only against pregnancy, no STI uses.
- Latex Condoms: Give the best prevention of STIs, but not the best prevention of pregnancy. (Failure rate for condoms is 10–20%!)
- Spermicidal compounds: Foam is best, since it kills not only sperm but also pathogens responsible for STIs—but it can also damage the cells in the lining of the vagina (making the woman more susceptible to STIs), and only lasts for ~1 hour. The failure rate for using spermicides without another method is 20–50%.

Fertility Awareness (rhythm method): Abstaining from sex on the days that the sperm might reach the egg. This is one of the least effective birth control methods, (failure rate is 20%) since it is very hard to calculate which days in the cycle are absolutely safe to have sex. The sperm can live 2–5 days in the female reproductive system (most only live ~28 hours, but five days is possible); oocytes only live 12–24 hours after ovulation, but it is hard to predict when ovulation occurs.

Withdrawal method: Also ineffective, (20–30% failure rate), since there can be sperm that escapes the penis before ejaculation (there is pre-ejaculatory semen before the urethral contraction occurs), and because it is also hard to ensure that the penis is removed in time.

Morning-After Pill: Can be used within the first few days after sex, and works in a variety of ways to delay ovulation, stop implantation, increase mucus production, etc.

Every year roughly 750,000 teenagers between the ages of 15 and 19 get pregnant in the United States. The Centers for Disease Control and Prevention estimate that there are 19 million new cases of sexually transmitted infections every year. So the bottom line is if you are sexually active BE SAFE, BE RESPECTFUL, and BE RESPONSIBLE. Hormonal contraception plus a condom is a safe combination for preventing both pregnancy and sexually transmitted infections. Other good ways to prevent sexually transmitted infections are to choose your partner carefully, communicate with one another, get vaccinated, and get tested and treated if necessary. Sex can be very beneficial (it builds bonds, burns calories, reduces stress, promotes a better night sleep for some, strengthens pelvic muscles, and the ultimate function, of course is to create another human being!), but be sure that you and your partner agree on the activity, are emotionally ready, and are both responsible for whatever the outcome.

Chapter 17

Genetics

DNA (deoxyribonucleic acid) is found in the nucleus of nearly every type of cell in our body. It is made from nucleic acids, which are made from nucleotides. Each nucleotide consists of a nitrogenous (nitrogen-containing) base, a pentose (5-carbon) sugar (deoxyribose), and a phosphate group. The nucleotides are covalently bonded together by the sugar and phosphate groups, and the nitrogenous bases are projected from the sugars and are free to bond with another nitrogenous base of a different nucleotide polymer.

- Watson and Crick identified the DNA structure to be two strands of nucleotides in a double helix, bonded together by hydrogen bonds between the nitrogenous bases.
  - o These bases always pair in a specific way, called the **Base pair rules**:
  - o Adenine always pairs with Thymine.
  - o Cytosine always pairs with Guanine.
  - o RNA (also made of nucleotide polymers) has no Thymine, but has Uracil instead, which bonds with Adenine.
- DNA: If one strand has the bases A-C-C-A-A-A-C-C-G, it will bond with

  T-G-G-T-T-T-G-G-C.
- When a cell divides (during mitosis), DNA is replicated identically. The two strands split apart like a zipper unzipping and new nucleotides are bonded in base pair rules to the old strands so that the two new DNA double helices now have one old strand and one new strand (this is called semi-conservative replication). Enzymes called **DNA polymerase** catalyze this replication by adding the new nucleotides to the old DNA strand, adding roughly 50 nucleotides per second in humans! (It occurs even faster in bacteria; about 500 nucleotides per second are added to the growing strands!)
- DNA cannot leave the nucleus (cannot penetrate the nuclear envelope). In order to get the nucleotide message into the cytoplasm of the cell, the DNA codes for RNA (Ribonucleic acid), which is a single strand of nucleotides.

- Remember the three types of RNA:
  - Ribosomal RNA (**rRNA**) helps to build ribosomes, which are the organelles necessary for protein manufacturing.
  - Messenger RNA (**mRNA**) takes the DNA message out of the cells' nucleus into the cells' cytoplasm where the message is **translated** at the **ribosomes** to create **proteins** out of amino acid chains.
  - Transfer RNA (**tRNA**) also helps with translation by bringing amino acids to the messenger RNA in the protein-building process.
- Messenger RNA (mRNA) can take the message out of the nucleus. If we look at the example above, the first strand of DNA would code for a messenger RNA strand that had the nucleotide sequence: U-G-G-U-U-U-G-G-C.
- DNA coding for RNA is called **transcription** (it is the same "language" of nucleotides despite the one difference in base pairs).
- This message goes to the ribosomes in the cell (remember, there are both free ribosomes floating in the cytoplasm as well as ribosomes attached to endoplasmic reticulum (called rough ER when ribosomes are attached).
- Ribosomes assemble proteins by translating the nucleotide sequence to code for an amino acid sequence. (Proteins are made from amino acid polymers.)
- Every three nucleotides code for one amino acid. A three-nucleotide unit is called a codon, and there are 64 codons in the human genome. There are only 20 amino acids, however, which means that many codons code for the same amino acid. For example, proline, an amino acid found in all organisms that have had their genetic code deciphered so far, can be assembled from the codons CCU, CCC, CCA, and CCG.
- In our example above, the mRNA UGG-UUU-GGC code for three amino acids Trp Phe Gly.
- Remember, proteins are polypeptide chains of amino acids in 3D structures, and both the sequence of amino acids and the shape specify the proteins' function. (See Chapter 2.)
- mRNA coding for proteins is called **translation** (since nucleotides and amino acids are different molecular "languages").
- The proteins that we make determine our traits: physical, chemical, behavioral characteristics.
- All instructions come from our DNA.

DNA inside our nucleus looks like a large mass of nucleotides and it is called chromatin in this state. When a cell goes through mitosis or meiosis, the chromatin condenses into discrete units called chromosomes. There are approximately 3.2 billion nucleotides in the DNA of every cell. The chromosomes range in length from 50 million to 250 million nucleotides. Humans have 23 pairs of chromosomes (46 chromosomes total) in all of our somatic cells (body cells). The paired condition of chromosomes makes these cells **diploid**.

There are no homologous pairs in gametes, which have only 23 chromosomes—these are called **haploid** cells, since they don't have chromosome pairs. Only 2% of the DNA in our chromosomes is in nucleotide sequences that code for proteins. These coding portions are called genes. Humans have approximately 25,000 genes. Although the Human Genome Project has identified all of the genes for humans, we still don't know what the functions of half of these genes are (we know they code for proteins, but we don't understand yet what the proteins do). These genes account only for the portion of the DNA that codes for proteins. Ninety-eight percent of our DNA does NOT code for proteins and is referred to as "junk DNA."

It was originally thought that this part of DNA had no function. But genetic research is now proving that this "junk" has very important roles to regulate the coding portions of DNA—like on and off switches to determine which genes will be expressed. For example, ancient fish had the genes that we have that stimulate the development of appendages; they just needed to be turned on and off at different times during development. A mutation in the regulation of these genes may have occurred at the time when amphibians started to develop so that they developed appendages, which allowed them to come out of the water. There is still so much to be discovered with genetic research!

Okay, now let's cover some basic genetics.

Gregor Mendel is considered the "father of modern genetics" due to his work in the Austrian Empire (now Czech Republic) in the 1800s. He discovered through monohybrid and dihybrid crosses of pea plants how alleles are expressed, and how they segregate, and independently assort. Modern genetics is still based on Mendel's laws of heredity from 1865.

Let's first start with some genetic vocabulary:

Genes: (classical definition) A unit of heredity responsible for controlling the formation of a particle trait/characteristic. Each chromosome contains many genes.

Gene (molecular): A strand of DNA that contains information coded in the sequence of nucleotides, which specify a series of amino acids that as a unit make up protein.

Genome: An organism's complete set of DNA. (Not all of the nucleotides in the DNA code for genes; in fact, only about two percent of the human genome is composed of genes coding for proteins. The rest is filler nucleotides providing support and regulatory functions.)

Alleles: Alternate forms of a particular gene, characteristically determining different phenotypes for the same trait (different nucleotide sequences code for different proteins that affect phenotypic expression).

Dominant: An allele that affects its phenotype in the heterozygous state (even if only one copy of the allele is present, it will determine the phenotype).

Recessive: An allele that can only express its phenotype in the homozygous state—two copies of the allele must be present.

Trait: Physical/biochemical or behavior character of an organism that appears to be genetically determined in a simple way; trait can refer to a specific phenotype.

Multiple Alleles: More than two possible alleles exist for a trait, but only two can occur in one individual for determining its specific phenotype (ex: blood type has three alleles, as we will discuss below).

Genotype: Which alleles you have determines your genotype. For example, if you have freckles, your genotype could be FF or Ff (if F signifies the freckle allele [dominant] and f signifies the non-freckle allele [recessive]).

Phenotype: Signifies how the alleles are expressed. In our example above, having freckles is the phenotype of both FF and Ff, and not having freckles is the phenotype of ff. You cannot always tell what the genotype is based on the phenotype, since there is often more than one genotype per phenotype (except for recessive phenotypes).

Incomplete Dominance: The dominant phenotype is not fully expressed; instead the heterozygous states give rise to a phenotype different from that of either allele in homozygous genes; ex: in snapdragons, crossing white and red flowers will result in pink flowers. Human skin color is also an example of incomplete dominance.

Co-dominance: When both alleles are expressed in the heterozygous state. An example is type AB blood. A and B alleles are both dominant over type O allele, but they are co-dominant with one another.

Monohybrid Cross: A breeding experiment that uses parental varieties differing in a single trait.

Dihybrid Cross: A breeding experiment where parental varieties differing in two traits are mated.

Polygenetic traits: There are numerous traits that are coded by the interactions of more than one gene. Eye color, height, skin color, and intelligence are all examples of polygenetic traits.

Epistasis: When alleles at one site affect alleles at other sites. An example is albinism. A person with the genotype for this disorder cannot make a particular enzyme, which in turn prevents the formation of the pigment melanin. Even though there are separate genes coding for hair color, eye color, and skin tone, if the person has the albino genotype, then they will lack pigmentation everywhere else as well.

Gregor Mendel's Laws of Genetics:
- **Law of Segregation**: Allele pairs segregate (separate) during gamete formation and the paired condition is restored by the random fusion of gametes at fertilization (by chance).
  - There are alternative forms for genes (alleles).
  - Each inherited characteristic has two alleles, one from each parent.
  - An egg or sperm carries only one allele for each inherited characteristic (haploid), since alleles separate from each other during the production of gametes (meiosis).
  - When two alleles of a pair are different, one is fully expressed and the other is completely masked (dominant/recessive).
- **Homozygous**: Identical alleles. There can be homozygous dominant (GG) or homozygous recessive (gg) genotypes.
- **Heterozygous**: Two different alleles (Gg).
- **Genotype**: Which alleles you receive (Gg, GG or gg are all examples of possible genotypes from the G/g alleles).

- **Phenotype**: The physical or behavioral expression of the trait. In our simplified example I will say that G is dominant and codes for green and g is recessive and codes for yellow. For both GG and Gg genotypes, the phenotype expressed is green. For the gg genotype the phenotype expressed would be yellow.
- **Rule of Multiplication**:
  - o The chance that an egg has g allele: 1/2.
  - o The chance that a sperm has g allele: 1/2.
- Overall chance two g's will come together: 1/2 x 1/2 = 1/4.
- **Punnett Square**: Green pods GG x yellow pods gg (recessive):

GG x gg

	G	G
g	Gg	Gg
g	Gg	Gg

GG x gg = all Gg (green)

Gg x Gg

	G	g
G	GG	Gg
g	Gg	gg

Gg x Gg = 3:1 phenotype (green: yellow)

Law of Independent Assortment: Each allele pair segregates independently during gamete formation.

Example: Parents have two traits: seed color and seed shape; rather than these traits being transmitted from parent to offspring as a package, each trait is inherited independently from the other.

- P = RRYY x rryy → all RrYy (R = round, r = wrinkled, Y = yellow, y = green);
- If four types of sperm mixed with four types of eggs, there are (4 x 4) 16 equally probable ways in which genes can combine;

Make up four phenotypic categories with ratios of 9:3:3:1 (9 yellow round, 3 green round, 3 yellow wrinkled, 1 green wrinkled):

	RY	Ry	rY	ry
RY	RRYY	RRYy	RrYY	RrYy
Ry	RRYy	RRyy	RrYy	Rryy
rY	RrYY	RrYy	rrYY	rrYy
ry	RrYy	Rryy	rrYy	rryy

- Some traits have intermediate inheritance (incomplete dominance), where F_1 hybrids have an appearance somewhere between the phenotypes of two parental varieties; ex: red snapdragons crossed with white yield pink.

- Genes are packaged into gametes in all possible allelic combinations as long as each gamete has one gene for each trait.

Multiple alleles: Some genes exist in more than two allelic forms.

- Human blood has three different alleles: I^A, I^B, (which are co-dominant) and i, which is recessive. Each individual still only gets two of these alleles (one from mom and one from dad) but there are 3 possible alleles in the population.
- There are 6 possible genotype combinations with these 3 alleles and they make the four different phenotypes for this trait (A, B, AB, or O).
- The blood type percentages in the US: 43% O, 40% A, 12% B, and 5% AB.

Antigens: Proteins on the outside of cells that are used as markers for the body to recognize self versus non-self cells. Foreign antigens elicit an immune response. In some cases antibodies will come to destroy the invader, or flag white blood cells to come destroy the cell with the foreign antigen.

- Blood cells are coated by antigens.
- Antigens are genetically determined by ABO alleles. (The type of antigen on blood cells is determined by which combination of the ABO alleles you have.)

Antibodies: Proteins we make in the spleen, by plasma white blood cells, to help combat foreign antigens. They bind to foreign antigens (like a lock and key) and form large insoluble complexes, which makes antigens harmless and facilitates the removal or destruction by our immune system cells.

- I^A allele encodes for A antigens on the surface of red blood cells.
- I^B allele encodes for B antigens on the surface of red blood cells.
- i allele is recessive to both I^A and I^B and encodes for neither A nor B antigens.
- Everyone inherits one of these alleles from each parent and which combination of alleles we receive determines our blood phenotype: A, B, AB or O.
- Which antigens you make will also determine which antibodies you make to combat foreign antigens.
- Type A blood can be from $I^A I^A$ or $I^A i$ genotypes. A person with type A blood will have A antigens and produce B antibodies to combat B antigens.
- Type B blood can be from $I^B I^B$ or $I^B i$ genotypes. A person with type B blood will have B antigens and produce A antibodies to combat A antigens.
- Type AB blood is from the genotype $I^A I^B$ These are co-dominant alleles and therefore they are both expressed. A person with AB blood will produce A and B antigens and therefore will not produce A or B antibodies.
- Type O blood is from ii genotype. A person with O blood does not have any antigens on the blood cells but can produce A and B antibodies to combat A and B antigens.
- Antibodies are produced if foreign antigens are encountered, for example in a blood transfusion with incompatible blood.
- A person with type A blood can only receive A or O blood.
- A person with type B blood can only receive B or O blood.
- A person with AB blood can receive AB, A, B, or O blood, and is therefore called a universal recipient. (This is because they do not make any antibodies to combat foreign antigens.)

- A person with O blood can only receive O blood since they can make both A and B antibodies. Since O blood does not have any antigens, however, it can be given to anyone, and is therefore called the universal donor.
- The best type of blood to donate is O-, since no antibodies will be made to fight off this type of blood, since there are no A or B antigens and no rhesus (Rh) antigens.

Rh Factor: Another blood group antigen on red blood cells that is separate from the ABO blood type alleles. For this factor there are only two types of alleles and the allele for Rh+ is dominant, and Rh- is recessive. This trait is named after the Rhesus monkeys, who also have this protein on their blood. Some of us have the Rh antigen and are + and some of us lack the Rh antigen and are -. This factor can cause problems for babies who have the protein but whose mothers do not, because the mothers will develop antibodies to fight the + Rh factor.

- If the mom is Rh negative, she lacks Rh antigens in her own red blood cells.
- If her fetus/baby is Rh positive, it must have inherited the Rh antigen from the father (because for the mom to be Rh negative she has 2 alleles that code for Rh negative)
- The mom will develop antibodies against the Rh antigens when a small amount of fetal blood crosses the placenta and comes into contact with her immune cells, usually late in pregnancy or at delivery.
- The first exposure is mild since the mom is just developing the antibodies so there are usually no consequences for this child.
- However if the mom has a second fetus who is Rh positive, the mother already has the capacity to make Rh+ antibodies (immunological memory), and these antibodies can cross the placenta during the final weeks of pregnancy to destroy the red blood cells of this second fetus.
- Pregnant women are monitored in these circumstances and can be given medication so that the mom won't develop her own Rh antibodies.

Sex-Linked Inheritance

- Genes are located on chromosomes, and chromosomes undergo segregation and independent assortment during meiosis to produce eggs and sperm.
- Sex chromosomes have a haploid number of chromosomes (23).
- Out of the 23 chromosomes in egg and sperm, one of these is a sex chromosome X or Y.
- Males are the **heterogametic** sex because their sex chromosomes in their gametes are not the same. They have XY in their somatic cells, therefore during meiosis 50% of the sperm will have the X chromosome, and 50% of the sperm will get the Y chromosome.
- Females are the **homogametic** sex meaning all eggs have X chromosomes (because females are XX in their somatic cells).
- **Sex-linked traits** refer to traits that are coded by genes that are located on the sex chromosomes. There are some traits from genes on the Y chromosome that only males get. Genes that are on the X chromosome can be expressed in both males and females depending on which combination of alleles are passed along.
- Since the X chromosome is much longer than the Y chromosome, there are many more X-linked traits. The X chromosome has 2,062 genes and Y only has 330, one of which determines "maleness"

and is called SRY. This gene codes for the formation of the testes in embryo, which then generate testosterone. If SRY is not present (in females), ovaries develop instead.

- Since most of the sex-linked genes are on the X chromosome, fathers pass sex-linked alleles to all of their daughters but to none of their sons. Since sons always receive their X chromosome from their mother, they can get X-linked traits from her but never from dad. (Sons always get the Y chromosome from dad.)

- Mothers can pass sex-linked traits to sons and daughters since X is the only sex chromosome mothers have in their eggs.

- Daughters will express only recessive X-linked traits if it is homozygous (if she received the trait from both mom and dad), whereas sons will always express it if they get the recessive trait on their X chromosome from their mom, since there is no corresponding gene on the Y chromosome. Because of this, there are more sons with these X-linked recessive traits.

- There are two X-linked recessive traits that we will cover here: **hemophilia** (not able to clot blood properly) and **colorblindness** (lacking the proper proteins to make all of the cones, which usually means being unable to detect red and green colors).

- When we talk about sex-linked traits we will always specify which alleles a person has as well as which chromosome the allele is on, since it is very important for determining the outcome of the offspring. This is different from the other punnett squares we have done where the genes are located on somatic chromosomes and we did not need to specify the chromosomes.

  o If a woman has one regular allele (H) that codes for the proteins that enable her to clot blood and one hemophilia allele (h) that does not code for the proper proteins to clot blood, she will still be able clot blood because she got one normal allele and can make the blood-clotting proteins. She will be called a "carrier" and her genotype is $X^H X^h$. (A "carrier" means that she has one gene that has the allele for the disorder, but since she doesn't have two, she doesn't express the trait because this is a recessive disorder.)

  o In order for a female to be a hemophiliac she would have to get two alleles that do not code for the blood clotting proteins—one on her mom's X chromosome in the egg and one on her dad's X chromosome in the sperm.

  o If she is producing offspring with a man who is a hemophiliac (he received the hemophiliac allele from his mother and no corresponding allele from his father—his genotype is $X^h Y$), the chances of them having a hemophiliac child is 50% (half of daughters will be $X^h X^h$ and half of sons will be $X^h Y$).

  o We could do the same example with colorblindness and get the same results. In fact, the hemophilia and colorblindness genes are located near each other on the X chromosome, so this is an example of traits that are often inherited together despite the rule of independent assortment. If the mother has a chromosome with both of the recessive alleles, 50% of her sons will most likely end up with both of those disorders.

	X^H	X^h
X^h	$X^H X^h$	$X^h X^h$
Y	$X^H Y$	$X^h Y$

- Most of the genes on Y chromosome have no homologous segment on the X chromosome.
- These genes are found only in males and are always passed down to sons.
- Examples are hairy pinna (long hair on outer ears) and the SRY gene coding for maleness (discussed above).

Genetics of Sex Determination
- Femaleness or maleness depends on which sex chromosomes you have—in humans XX = female and XY = male.
- The sex is determined by the type of sperm cell that fertilizes an egg (remember, sperm cells are heterogametic so can be either X or Y).
- Random fertilization results in a 1:1 ratio.
- Y determines maleness due to the SRY gene on the Y chromosome.
- XX does not determine femaleness, because some males have XXY.

Incorrect chromosome number can occur if chromosome pairs fail to separate correctly during meiosis. Often the second stage of meiosis does not separate correctly in older oocytes, leading to eggs with too many or too few chromosomes, or defective sperm are produced with the wrong number of chromosomes. If one of these abnormal gametes is involved in fertilization, it can lead to an individual with **trisomy** (one extra chromosome) or **monosomy** (one less chromosome), and therefore have more or fewer than the normal 46 chromosomes. Many trisomys and monosomys are spontaneously aborted (miscarriages are often due to chromosomal abnormalities), but one trisomy that is not usually spontaneously aborted is trisomy 21 (there are 3 chromosomes in the twenty-first spot rather than the regular pair). This is known as Down syndrome. There are many different symptoms associated with Down syndrome, ranging from mild to severe. Trisomy 21 can affect physical, mental, and social development, and is the most common cause of birth defects. Pregnant women over thirty-six often undergo an amniocentesis in order to find out the karyotype of the fetus in order to check for trisomy 21, trisomy 18 (Edward syndrome), and Turner's syndrome (XO), since these are some of the chromosomal abnormalities that may occur in later-in-life pregnancies.

Turner's Syndrome has to do with the sex chromosomes in the fact that the individual only gets one sex chromosome—an X. The genotype of the individual is portrayed as XO (only one X chromosome) and this is a female (since there is no Y). Some women with Turner's syndrome are mentally retarded and sterile, but many have only slight physical limitations and limited fertility. XXX is also a possibility (also female) and most develop normally.

- **Other cases of too many sex chromosomes that lead to viable individuals:**
- **Klinefelter Syndrome** is the term for males with an extra X chromosome (XXY). A person with this genotype still has male sex organs, but some have abnormally small testes, enlarged breasts, and mental deficiencies; some are sterile.
- XXXY (or multiples of Xs) are still males with similar symptoms as Klinefelters. There are also XYY males, some of whom have mild mental impairment.

Other Sex-Associated Genetics:

- There are **sex-limited genes** that both men and women have, but are expressed differently depending on how hormones regulate them. For example, there are genes on the X chromosome that code for breasts and beards, but the development of these traits is controlled by estrogen and testosterone, so generally women end up with breasts and men have beards. Hormone imbalances can result in women with beards and men with breasts, since we all have the genes for these traits. This is different from a **sex-influenced trait**, which results in different genotypes coding for the same phenotype depending on the gender. For example, in males the phenotype for baldness is expressed in the heterozygous state, but for females the baldness phenotype is expressed only in the homozygous recessive form (relatively rare).
- **Genetic Disorders**: There are more than 15,000 genetic disorders from missing or malfunctioning genes, which can have mild, to severe, to fatal effects. A gene mutation can occur when pieces of chromosomes are improperly duplicated, deleted, or shuffled around, so that the sequence of nucleotides no longer codes for the functional gene. A part of one chromosome may even change places with a part of another chromosome that is not its homologous pair. This is called translocation. Mutations can be beneficial, have no effect, or can be deleterious. Gene mutations can then be passed along, since the new form of DNA will be duplicated during meiosis and incorporated in the gametes. (There are also mutations that are not inherited, such as mutations that cause cancer.)
- Mapping the human genome (started in 1990 by scientists from around the world and finished in 2003) has greatly enhanced our understanding of all of the genes that humans have, as well as some of the genetic disorders. James Watson (noted for co-discovering the structure of DNA with Francis Crick and the head of the Human Genome Project for a couple of years) was the first person to have his DNA mapped. Now anyone who is willing to pay the price (hundreds to thousands of dollars, depending on the level of the analysis) can send in a sample of saliva (which contains cheek cells with DNA) in order to have their genetic make-up sequenced. It's quite amazing.
- One of my students shared the results of her DNA test and there were so many traits that I didn't even know about. The test gave her specific trait results, disease risks, carrier status, and family history information. I learned that early menopause is a genetic trait and that you can test for "eating behavior," "HIV resistance" "longevity," "memory," "smoking behavior," "malaria resistance," and "muscle performance" (she's an "unlikely sprinter"!). Yes, these were all traits tested for from the DNA of her cheek cells. I find it fascinating. There were many different disorders listed as possibilities along with her risk assessment for these disorders.
- Many disorders have to do with environmental factors as well as genetics (for example, drug use, nutrition, and exercise all have roles in how some genes will be expressed). If you are at risk for diabetes II and start to lose weight, exercise, and maintain a healthy diet, you may be able to avoid the disease before you even see any symptoms. This is an incredibly useful way to use genetic profiling. There are

also genes that can tell us about how people will react to different medications, whether they will be receptive to warfarin, for example (the medication given after a heart attack). This is valuable information to have and as genetic profiling becomes more prevalent, it will really change and enhance the health industry. We need to move toward more preventative care as our life expectancy increases (the U.S. average is now 77 years old). We pay far more in healthcare costs than other countries with the same life expectancy averages because of our lifestyles and the fact that we don't focus on preventative care as much as we should.

Genetic Disorders

There are many different types of genetic disorders but I want to give some of the classic examples for this text:

There are three different types of **genetic disorders**:

1. Autosomal recessive
2. Autosomal dominant
3. Sex-linked.

Autosomal means that the disorder is coded by alleles on the somatic chromosome (located on a pair of the 22 chromosomes that are <u>not</u> the sex chromosomes).

Autosomal recessive disorders occur if the individual receives two of the same recessive alleles, one from mom and one from dad.

- <u>Sickle cell anemia</u>: Blood cells are a sickle shape due to a problem in the hemoglobin protein. The sickle-shaped blood has a harder time traveling through the blood vessels and they also have a shorter life span. The spleen works hard to try to rid the body of these sickle blood cells, and the person is anemic with not enough healthy blood cells or oxygen. Spleen damage and clots in capillaries all result from sickle cells, and these can lead to lung, brain, and heart damage or failure, and infections, paralysis, kidney failure, and death. In the heterozygous form (one sickle cell gene and one normal gene), the trait that results is actually beneficial, since people with this genotype do not have sickle cell anemia and are also less likely to contract malaria. The disease is more common in black people around the world (8%); only 1% of Latinos have this disorder. It is relatively rare in other races. When one gene causes multiple phenotypic effects it is called <u>pleiotropy</u>.
- <u>Albinism</u>: In the recessive genotype, the person cannot make any of the pigment melanin. So skin, irises, and hair are all white. (This is an example of epistasis, which we discussed earlier in the chapter.)
- <u>Cystic fibrosis</u>: Leads to thick mucus production in liver and lungs or intestinal tract. Person is more susceptible to bacterial infections. It is the most common of the lethal genetic disorders in the United States.

- <u>Tay-Sachs</u>: A lack of the enzyme that helps metabolize a certain lipid, so it ends up accumulating in the brain, causing damage to the brain.

Autosomal dominant disorders require only one dominant allele to display the disorder (the allele can be passed on from mom or dad). They are less common because in many cases where there is a defective dominant allele, the zygote, embryo, or fetus will be spontaneously aborted. There are a few dominant disorders, however, that are not spontaneously aborted:

- <u>Huntington's Disease</u> is a nervous system disorder where the neurons deteriorate and motor control fails over time. Most people with this disorder do not even know that they have it until after 40 when the symptoms begin. Although if one parent has the disorder, it is now recommended to get genetic testing for this disorder.
- <u>Dwarfism</u> is the heterozygous state of a chondroplasia. In the homozygous dominant form, the fetus is stillborn. The homozygous recessive genotype codes for normal growth.

Sex-Linked disorders have already been discussed. Examples are <u>color blindness</u> and <u>hemophilia</u>.

- Hopefully, with more genetic research in the future, we will be able to treat some of the genetic disorders that have been identified by inserting normal copies of the gene needed to make particular proteins. Gene therapy is a growing field, but at this stage it is very tricky to do and doesn't always work. We will also be able to assess risk, modify behavior, and know which medications will treat diseases more efficiently. It will be a very exciting field to follow.

Chapter 18

Cancer

Cancer is the second leading cause of death in the United States (after heart disease). It is basically an acquired genetic disease since it is not inherited from birth, (although some genes can make people more prone to developing cancer). It develops through a series of steps in which DNA changes (mutates) due to different causes such as tobacco (#1 threat), or other carcinogens, poor diet (obesity is the #2 threat), UV radiation, viral infections, or random mutations to the DNA. New studies have shown that 40% of cancer cases could be avoided by living healthier lives (not smoking, eating lots of fruits and veggies, maintaining a healthy BMI, exercising, and limiting alcohol consumption). We will be discussing the main risks for developing cancer and ways to avoid this horrible disease, but first let's look at what cancer is and why it develops.

DNA controls cell division (mitosis) by producing proteins that regulate checkpoints in mitosis. These checkpoints stop cell division from occurring if something goes wrong with replication or division, and also stops us from making too many cells if an area is getting overcrowded with cells. We also have built-in indicators on our chromosomes to stop cells from dividing after they have gone through mitosis a certain number of times (usually 50–60). Cancer is due to different mutations in the DNA causing problems with the proteins that regulate cell division, and therefore <u>cells divide out of control</u>. This is what cancer is—out-of-control growth in mutant (cancerous) cells that start to take over different parts of the body, causing the normal cells to die. Many cancer cells will continue dividing indefinitely (until they are killed or until they kill the person they have spread inside of). Immortal, out-of-control, abnormal cells in the human body is a deadly combination.

Cancer occurs in one in every three people in the U.S., and kills one out of four—this country sees 1,500 deaths every day, and more than half a million/year (statistics are from 2002; there may be even more now). The number-one cause of cancer is tobacco and the number-two cause is obesity and not enough fruits and vegetables. Cancer affects more males than females (probably due to lifestyle choices). There are more than 100 types of cancer but the top four are prostate, breast, lung, and colon. Prostate is the number-one cancer in males, breast is the number-one cancer in females, but both sexes **die** the most from lung cancer.

Americans develop breast and prostate cancers 5 to 6 times more than people of some other countries, most likely due to lifestyle choices (diet, alcohol, and exercise really affect these glandular cancers). We can reduce our risks with the choices we make about how we live our lives.

You don't inherit cancer and are not usually born with cancer (although there are instances of babies born with aggressive stages of cancer). It is not a hereditary genetic disease, although some people inherit genes that make them more prone to developing cancer. For example, we have genes that suppress tumor growth called Tumor Suppressor Genes. These genes halt cell growth and division, thus preventing tumors from developing, and they can prevent oncogenes from being expressed. If you inherit mutations in these genes, it can make you more prone to developing cancer. Inherited susceptibility to cancer accounts for only about five percent of cancer cases, and is mostly associated with breast, lung, and colorectal cancers. (Ex: Women who inherit mutated versions of the BRCA1 and/or BRCA2 genes are more likely to develop breast cancer. Of the 200,000 women a year diagnosed with breast cancer, five to ten percent of them have one or both of these gene mutations.) Men can inherit BRCA genes as well and develop breast cancer; it is just not as common as in women.

Cancer starts with **proto-oncogenes**—genes that code for proteins that regulate cell division in some way. Proto-oncogenes can turn into **oncogenes**—genes that do not respond to the control signals that regulate cell division. At least three percent of our 25,000 genes are proto-oncogenes. Any of these can trigger cancer if they mutate.

The Main Causes of Cancer

The main reason for cancer is mutation in the DNA of somatic cells. There are many reasons mutations may occur, such as the different types of carcinogens, radiation, viruses, or just random mutation. Our body is constantly fighting off these mutations, but there are instances when we can't fight them as efficiently. After the main causes of cancer, I have also listed reasons people are more likely to succumb to cancer.

Carcinogens are cancer-producing molecules with free radicals that can also cause changes to DNA. (Free radicals are molecules that are unstable due to unpaired electrons in the outer valence shell. This instability causes the free radicals to strip electrons from other molecules, which can change the composition and function of the affected molecule. We need a certain amount of free radicals for different biological processes to occur, but excess free radicals can cause cell damage and death. Damage to cells due to free radicals is linked to many disorders, not just cancer. In some cases they are linked to heart disease, stroke, diabetes, Parkinson's, schizophrenia, and Alzheimer's.) You can neutralize these free radicals by consuming foods and drinks that contain antioxidants. Antioxidants are chemicals that stop oxidation from occurring. (Oxidation refers to the loss of electrons from molecules, atoms, or ions). Antioxidants stop free radicals from stripping electrons away by donating electrons to the free radicals, thus neutralizing the free radicals. Vitamins A, C, and E, and flavonoids (used to be referred to as vitamin P) are all antioxidants. A diet rich in these antioxidants is very beneficial for reducing free radicals and their associated risks. They are found in black and green tea, red wine, dark chocolate, and numerous fruits and vegetables, such as leafy greens, blueberries, cranberries, oranges, apples, and pomegranates. For an example of how potent different drinks are with antioxidants: 1 glass of red wine is equivalent to 2 cups of black tea, 7 glasses of orange juice, or 20 glasses of apple juice!

Types of Carcinogens:

There are many types of carcinogens but here are some of the main ones: hydrocarbons in tobacco (80–90% of lung cancer cases are due to tobacco), pesticides, diesel exhaust, benzene (in paints and dyes), asbestos, and nitrate preservatives in processed meats (like hot dogs and sandwich meats). Other foods that can contain mutagenic chemicals are charred BBQ and smoked or pickled foods. (Occasionally enjoying these foods is fine, just don't have them too often.) Soot from chimneys can be carcinogenic (many chimney sweeps used to get lung and testicular cancers), and chemicals leaching from plastics can cause gland cancers like breast and prostate. Some plants and fungi produce carcinogens as well.

As mentioned above, carcinogens are not the only causes for cancer—here are the other main reasons people develop cancerous cells:

Radiation: UV, X-rays, and gamma rays can all lead to DNA mutations (cytosine and thymine are particularly vulnerable to UV radiation, which causes the three types of skin cancers). Even radiation from cell phones can cause cancer. (Don't keep them in your bra or pants pockets.)

Viruses: Some viruses carry oncogenes in their DNA and can insert these genes into human DNA and cause cancer by altering the host DNA. Examples of viruses that cause cancer are HPV (causes mostly cervical and ovarian cancers) and Hep B (causes cancer of the liver).

Random mutation in DNA during cell division of somatic cells sometime occurs as well. For example, one form of leukemia (white blood cell cancer) is due to part of chromosome twenty-two attaching to chromosome number nine, causing gene mutations due to translocation. In other instances it can be a smaller mutation where just a small portion of the nucleotide sequence on a chromosome is shifted, resulting in oncogenes.

Risk factors for developing cancer:

Smoking is the #1 risk factor for developing cancer. Exposure to other types or carcinogens can put you at risk as well. For example, lung cancer is also found frequently in farm workers who are exposed to pesticides daily. People exposed to benzene (working with paints and dyes a lot) are more at risk for leukemia. Try to cut down your exposure to different types of carcinogens. Exposure to UV radiation and viruses will also increase the risks of cancer, so protect yourself from the Sun and wash your hands frequently, get vaccinated, and practice safe sex to protect yourself from viruses.

Diet, weight, physical and mental health, and hormone levels can influence the risk of cancer as well. A diet low in fruits and vegetables means fewer antioxidants available to help neutralize free radicals. High-fat and -protein diets are more likely to cause certain types of cancers because they cause increased inflammation, which helps to feed cancer growth. Diets with a lot of sucrose and high fructose corn syrups will also help to feed cancerous growth. (Diet problems account for approximately thirty percent of environmentally caused cancers.) Increased adipose tissue is especially associated with breast cancer in women because it increases estrogen levels, which triggers more cell division in the breast tissue. (Higher estrogen levels can also be due to never having a child, never breastfeeding, and having long-term hormone treatment either for birth control or replacement therapy for menopause.) Alcohol abuse can also increase the risk of cancer since it suppresses the immune system and feeds cancerous growth. Diet, physical fitness, mental health, and hormone levels all affect the immune system, so if one or more of these factors is unbalanced, it is more likely to lead to a breakdown in immunity.

Breakdown in immunity can cause an increased risk of cancer. Our immune system is constantly fighting abnormal cells (we kill off about two tumors a day with our immune system). If we are not able to fight off these precancerous cells, they will start to divide and grow into cancerous tumors. (Our WBCs, such as T-cells and NK cells, detect and kill any abnormal precancerous cells, and if our immune system is compromised, then cancer cells are more likely to grow and spread.) Compromised immunity can be due to stress, depression, autoimmune disease, HIV, chronic respiratory ailments, and being on immunosuppressants for organ transplants. You can prevent certain types of cancer by getting vaccinated and practicing safe (condom) sex. The Hep B and HPV vaccines can help to avoid getting cancers such as liver, cervical, and ovarian. Getting enough sleep is also critical to keep your immune system healthy. Maintaining a positive mental outlook on life will also help. (Being constantly stressed out and having negative feelings trigger the secretion of more adrenal hormones, which often causes increased inflammation, which can feed cancerous growth.)

Now let's look at what the differences are between normal and cancerous cells:

Normal Cells	Cancerous Cells
Controlled growth	Uncontrolled growth due to mutations in DNA, which stop your cells from regulating mitosis (proteins coded for by DNA are important to stop mitosis, and if you can't make these particular proteins due to the mutation in the DNA, your cells can grow out of control)
Contact proteins regulate spacing, recognition, and cohesion	No contact proteins, so there is no regulation of spacing, which leads to cells piling on top of each other as they continually divide. There is also no cohesion, which allows the cancerous cells to break free from the pile and move to other areas since nothing is adhering them together
Uniform Layer	Non-uniform layer: cells pile on top of one another
Differentiated (cells have specific functions)	Undifferentiated cells, no specific functions
Smooth and no pseudopodia	Ragged edges and pseudopodia extensions ("fake feet"), which help them break free from piles and move around (in circulatory or lymphatic systems)
Normal Nucleus/Cytoplasm ratio	Nucleus is large/cytoplasm small; chromosomes and DNA larger*

*Karyotyping cancer cells can show the problems with chromosomes: problems with replication lead to more chromosomes (trisomy and even four or more of the same chromosome can happen), some parts of chromosomes are duplicated more than they should be, and some parts are deleted so pairs often don't look homologous any more, since one is often longer than the other, etc.

There are many different stages of cancerous growth. When cells start rapidly dividing it is called **hyperplasia**. Once these rapidly dividing cells change form and look like typical cancerous cells, it is called **dysplasia**. A mass of these cells is called a **tumor**. If the tumor is contained in one area and is not spreading, it is called a **benign tumor** and its remaining in place is referred to as "**cancer in situ**." Usually benign tumors are enclosed in a connective tissue capsule. If a tumor starts to grow (if the cells break out of the capsule), it is called a **malignant** or **cancerous tumor**. If the immune system doesn't kill these cells (remember, our immune system is fighting off abnormal cells every day), and the cancerous cells start to spread to another area, it is called **metastasis**. For cancer cells to grow and move, they need their own blood supply.

The cancerous cells start secreting growth-enhancing proteins and chemicals called **angiogenic compounds** that attract a blood vessel to come over to the tumor. The cancerous growth needs blood for nutrients as well as to get rid of wastes as it divides. Cancer cells can travel in blood vessels or lymph vessels. They secrete digestive enzymes in order to get into a vessel. Then the cancer cells can travel throughout the body (using their pseudopodia to tumble along) and squeeze out of the vessels into other tissues to start new tumors. These angiogenic compounds increase inflammation in the area, which helps to feed the cancerous growth. (The poisonous chemicals secreted into the surrounding tissue by the cancer cells trigger the immune system to fight back by secreting cytokines from white blood cells, which causes inflammation. The extra dilated blood vessels in the area will bring more nutrients to the cancer cells and meanwhile the mutant cells will deactivate the white blood cells that have come to fight the infection.) High-sugar and -fat diets also increase inflammation. Obese people have much more inflammation than people with healthy body weights. (They also produce more estrogen from their adipose tissue, which is a risk factor for hormone-sensitive cancers like breast and prostate.)

People who take anti-inflammatories are sometimes at a lesser risk for cancer, but too much anti-inflammatory medication can lead to other problems such as ulcers and gastritis. There are natural anti-inflammatories, such as catechins in green tea and resveratrol in red wine, to help stop cancer from secreting proinflammatory substances. A healthy diet in general and exercising will help reduce inflammation. Some cancers are related to chronic inflammatory sites like the liver (usually caused by the hepatitis virus), stomach cancers (often from helicobacter pylori bacteria), or cervical cancer (from the papillomavirus.)

The colon often becomes inflamed as well and is a major area for cancerous growth. There are 50,000 deaths from colon cancer per year. Many of these could have been prevented with cancer screenings and healthier lifestyles. Colon cancer is a very slow-growing cancer with polyps growing for 10–30 years before they really start to have an impact on the body. (A **polyp** is an abnormal growth projecting from a mucus membrane.) If a polyp is detected on a colonoscopy (recommended every year after the age of 50), it can be removed right then. If the polyp turns into a tumor, it can progress very quickly into a deadly cancerous growth. It is one of the most preventable cancers but not enough people get screened yearly. Drinking milk, limiting red meats to 18 ounces or less a week, exercising, and flushing out the colon regularly (maintaining a healthy high fiber diet—at least 28 grams per day) will reduce the chances of developing polyps. Tobacco and excess alcohol are also risk factors for colon cancer, so limiting these will help as well.

Often cancer deaths are not from the original cancer site but are from where the cancer moves to. Often the first main capillary bed that metastisized cells travel to is in the lungs. They exit the capillaries there and cause lung cancer in the epithelial tissue inside the tubules. (Remember, out of all of the different types of cancers, lung cancer causes the most deaths in both males and females.) Another place cancer cells will start new growth is the brain, causing tumors in glial cells or in meninges. Brain cancer is very hard to treat and can cause death very quickly.

Cancer cells grow out of control due to lack of proteins that stop cell division, but they also can live forever because they have a protein that normal cells don't usually have—it constantly repairs DNA so that the cell won't die. Normal cells usually divide 50–60 times, and then they die. Every day, 50–70 billion cells die this way in an adult. This natural cell death is called apoptosis, and is very different from when cells die due to traumatic cell death from some sort of damage that occurs before the end of their natural life span (called necrosis). Parts of our chromosomes, at the tips, are called **telomeres**, which get snipped off every time cells divide, and once the telomeres are all gone, the cells stop dividing and the cell dies. It does not spill out its contents, it just shrivels and immune system cells come clean up the debris. Cancerous cells often have a mutation in DNA, which makes an enzyme called telomerase, which repairs the telomere sections that are snipped off at every division so the cells never stop dividing. In fact, there are some cervical cancerous cells that were taken from a woman named Henrietta Lacks in 1951 that are still dividing (called HeLa cells). They have been used for research on cancer, vaccinations (they helped develop the polio vaccine), AIDS, and gene mapping for the past 60+ years! Unfortunately the family of Henrietta Lacks was not told that her cells were harvested, developed, and sold, until the 1970s, when researchers started contacting them for more information about their genetics. (A very interesting story that has been documented in a book and movie.) Hopefully these cells will help to save more lives in the future as we learn more about preventing and treating cancer.

Major Types of Cancers

Sarcoma: Cancer in connective tissue, such as muscle and bone
Carcinoma: Epithelial cancers, skin (see Chapter 5), and organ linings, like lungs
Lymphoma: Cancer in lymph system
Adenocarcinoma: Cancers in glands, such as breast and prostate
Leukemia: Bone-forming regions, abnormal increase in stem cells, WBCs, platelets, or RBCs
Glioma: Cancer of the brain (develops from glial cells or meninges)

Detecting Cancer: Warning signs spell CAUTION (from the American Cancer Society)

C	changes in bowel or bladder habits
A	a sore that persists
U	unusual bleeding or discharge
T	thickening or lumpiness of tissue
I	indigestion or problems swallowing
O	obvious growth in a wart or mole
N	nagging cough or sore throat

Some cancers you can **detect by yourself**: checking for lumps or noticing changes in your health, such as from the list above. There are other ways that cancer can be detected as well:

MRI, X-rays, ultrasounds, radioactive tracers, and measuring the levels of certain substances ("**markers**") **in the blood**. For example, the human Chorionic Gonadotropin hormone (hCG), which is usually produced by a developing embryo in order to maintain the corpus luteum during pregnancy, can also be produced by some kinds of cancer cells. If there is hCG in the blood and it's not due to pregnancy, it could signal that cancer is occurring somewhere in the body. Other markers include prostate-specific antigen (PSA), as well as an ovarian cancer marker. The most reliable way to confirm a diagnosis of cancer, however, is to do a **biopsy**. In this method a sample of tissue where cancer is suspected is removed by surgery or by a hollow needle and then examined with a microscope to see if cancerous cells are present. Women should get their pelvic exam (PAP smear) done regularly to check for cervical cancer since it is one of the deadliest cancers for women after breast cancer.

Reducing the Risks of Cancer:

Avoid tobacco and other carcinogens.

Get checked regularly (especially breasts, cervix, prostate, and colon).

The HPV **vaccinations** guard against many types of cancers. For women HPV causes cervical and ovarian cancers. Recent studies show that males are getting oral, head, and neck cancers due to HPV as well. Hepatitis vaccinations help to guard against liver cancer.

Exercise and maintain a **healthy weight** (in women, adipose tissue secretes more estrogen, which can lead to increased risk of breast cancer—exercise decreases estrogen levels). Be careful with hormone medication for this reason as well (estrogen replacement after menopause is linked to breast cancer as well).

In everyone, having excess weight causes excess inflammation, which feeds cancerous growth, so again … maintain a healthy weight; this will also help with having a positive, self-confident attitude.

A **healthy emotional state** with confident, positive attitudes and relationships will also promote strong immune systems. (Depression, anger, and social isolation all inhibit the immune system.)

Proper nutrition with plenty of **antioxidants** to help neutralize free radicals, and plenty of anti-angiogenic foods to help block effects of angiogenin and therefore stop the blood vessels from feeding the cancerous tumors. (Some anti-angiogenic foods are green tea, red wine, kale, bok choy, ginseng, licorice, strawberries, blackberries, blueberries, raspberries, oranges, soy, grapes, grapefruit, lemons, apples, pineapples, pumpkin, tuna, garlic, tomato, olive oil, dark chocolate, artichokes, Earl Grey tea, and parsley.) (Mediterranean, Indian, and Asian diets are all more anti-inflammatory than western diets, which are high in sugars and fats, both of which increase inflammation.)

Exercise and proper nutrition also help to keep your immune system strong, and since that is your first line of defense, that is very important!

Limit alcohol consumption (breast cancer risks are higher for women who drink more than five drinks/week). Also, if your body is always working hard to detoxify your blood from alcohol, it won't be as effective at fighting off infections and precancerous cells. The exception to the alcohol rule is red wine (in moderation), which is actually beneficial to fighting off cancer.

Avoid high UV exposure—remember to wear sunscreen and a hat!

Avoid other radiation exposure (don't carry your cell phone in breast pocket or pants pocket).

Avoid stress and **get enough sleep** to keep your immune system functioning properly.

Practice safe sex (viral infections can lead to cancer).

Treatments for cancer are chemotherapy (drugs to kill cancer cells), surgery (extract the tumor if it is contained), radiation (kills both cancer cells and healthy cells in an area), and anti-angiogenic therapy (drugs to control the angiogenesis that occurs: if you can stop the blood vessels from feeding the cancer cells, you may be able to stop the growth). For leukemia, patients are treated with chemotherapy to suppress DNA synthesis in the affected white blood cells, followed by a bone marrow transplant so that the new stem cells will produce healthy white blood cells.

Chapter 19

Evolution, Ecology, and Conservation

Evolution means adaptive genetic change over time and gives rise to diversity on many different levels: populations, species, and ecosystems. Charles Darwin was the first biologist to formulate a scientific argument for the theory of evolution, based on two aspects that work together to cause evolution—natural selection and adaptation. He developed his theory from data he collected while studying the organisms and geography on a 5-year voyage from 1831 to 1836 aboard the *H.M.S Beagle* as a naturalist. They traveled from England to South America, where they sailed along the eastern coast, around the tip, and up to the Galapagos Islands. (Darwin made the finches of the Galapagos eternally famous.) The *H.M.S. Beagle* then sailed across to New Zealand, Australia, Africa, back to South America, and then returned home to Plymouth. Darwin didn't publish his theory until 1858 when he and Alfred Russel Wallace presented their scientific papers to the Linnean Society of London. His theory became more widespread when he published his book <u>On the Origin of Species</u> in 1859.

The theory of evolution has been in place for more than 150 years now. A scientific theory remains valid until it is proven wrong, and despite many efforts on the part of the non-believers, no one has been able to disprove evolution. It is still a hotly debated topic, however, because some people refuse to believe that humans could possibly have evolved from primitive organisms. There is scientific evidence everywhere to support the evolution of life. The evidence is found in historical observations, the fossil record, and molecular data. We all evolved from primordial life in the sea and we're still evolving.

Let's look at some of the observations that led to Darwin's historical theory. Darwin noticed many trends that he found puzzling. Why did South America have unique animals not found anywhere else? Organisms on different islands seem to have many similarities (common forms) but are different species, suggesting common ancestry. He noticed that marine organisms on either side of Panama were very different from one another despite being only a few miles apart. He was also curious as to why there are organs in some animals that appear to have no purpose but resemble organs that do have a specific function in other species, and that many species have "homologous" structures, meaning structures that are similar in form and function, despite not being similar organisms. (The appendages of humans, cats, bats, and whales are all homologous structures despite how different we are from one another.) Another main realization for Darwin was that all vertebrate embryos have similarities that suggest common ancestry.

These observations led to Darwin's **Natural Selection Theory**, which is basically that the organisms that are "most fit" (best adapted to their environment and able to find food and mates) will survive more successfully and therefore pass along more of their genes and traits to the next generation. The population will evolve due to this adaptive shift in genes. Individuals do not evolve; rather the change is very gradual over generations.

Five main deductions that led to the Natural Selection Theory:
- Most features are determined by heredity.
- There are heritable variations within populations.
- Variation can lead to different rates of reproductive success and survival.
- Differential reproduction fitness and survival leads to shifts of characteristics within populations.
- If this process of differentiation continues long enough, speciation can occur (where parent and daughter populations can no longer produce fertile offspring).

Evolution gives rise to diversity, and on the flip side, variation provides the opportunity for evolutionary change and can speed it along. Starting on the molecular level, **genetic diversity** is due to mutations in the DNA (which can be advantageous, benign, or deleterious). It is also caused by sexual reproduction, which results in the shuffling and mixing of genes due to recombination of DNA during meiosis, as well as the shuffling of genes depending on which egg and sperm unite. Genetic diversity can lead to population diversity, which then can lead to speciation and therefore species and ecosystem diversity as well. **Speciation** is when two populations within a species diverge so much from one another that they can no longer reproduce sexually to produce fertile offspring. This is also known as reproductive isolation, and the populations become two separate species. (Remember that the definition of a **species** is that it is the smallest unit that can interbreed and produce fertile offspring. Although some species can be crossed, they usually don't produce offspring that can reproduce; for example, a lion and tiger can be crossed, but the resulting "liger" couldn't produce another liger.)

Evolution is moving from primitive species to more complex species. In order to study the evolutionary history of organisms, biologists look at fossil remains to identify which organisms lived where and when. Fossils found at the lowest levels of rock are the most primitive, and those found in higher elevations of the sedimentation tend to be more advanced. **Fossils** are evidence or physical remains of organisms that lived in the past. Examples of fossils are skeletons, teeth, shells, seeds, and imprints in rocks (like tracks). Biologists not only look at individual species (fossils or living), but they also look at relationships between species and relationships between species and their environments—studying these relationships is called **Ecology**. Studying both ecology and evolution help us to better understand the world and is crucial for **conservation**, the study of species, habitats, and ecosystems with the goal to protect and conserve nature's biodiversity.

There are **three main levels of biodiversity** important for evolution:

1. **Genetic diversity**: (already mentioned) Everyone, except identical twins and clones, has different DNA (and even twins have different fingerprints and iris patterns, so they are not really identical).

2. **Species diversity**: (huge) Two million have been named, but estimates go up to 100 million. Scientists discover roughly 15,000 new species per year! A recent mammal discovery was a bat the size of a raspberry, but most of the new discoveries are not mammals. There are many invertebrates, plants, fungi, protistas, and bacteria discovered all the time, however. We've already talked in Chapter 1 about some of the roles different species have within their communities. Some are producers, making oxygen and foods that the rest of the organisms need to survive; some are decomposers, breaking down organic materials and returning the nutrients to the soil. There are numerous consumers that have their own functions, such as regulating other species' population growth as well as other roles within the community. Another example of a group of extremely important consumers is the pollinators, which help to spread the angiosperms (flowering plants).

You may not know what every species' function is in this world, but just know that they all have a role in their community. Whether they are pollinators, pest control, population control, or food for others, they are important. Some species are more critical to their communities than others, since they affect more of the species around them, and therefore are called "**keystone species**." For example, sea otters are keystone species in their ecosystem. If they go extinct, there will be no one to control the population of sea urchins (sea otters eat sea urchins), and the sea urchins will consume much of the kelp in the ecosystem, which provides food and protection for multiple other species. It's a domino effect; when a keystone species is removed from the community, many other species are adversely affected. Some ecosystems are extremely important for the world at large and have a similar domino effect when they are destroyed.

3. **Ecosystem diversity**: An ecosystem includes the species that live in an area as well as the abiotic factors like water, air, rocks, and atmosphere. Examples of ecosystems are tundra, rain forest, desert, temperate regions, tropical regions, and wetlands. Each ecosystem has important functions, just as each species plays a role in the web of life. Forests are critical ecosystems, so I will list some of their functions to give you an example of the role of ecosystems and how it relates to us.

Functions of forest ecosystems

1. Climate control: Forests help to absorb CO_2 as well as absorb and release water, all of which help to cool things down.
2. Plant communities in forests help control soil erosion; fertile soil absorbs water, which packs down the earth.
3. Controls flooding, since the soil is absorbing the water.
4. Controls sediment buildup in lakes, rivers, and reservoirs, due to the control of erosion.
5. Home to so many species (80% of terrestrial organisms live in forests!).
6. Produces a lot of oxygen and uses CO_2.
7. Provide wood for fuel, shelter, and other goods.
8. Forests provide more than half of all of our medications.

Results of deforestation (when forest ecosystems are destroyed)

1. Increases air temperature due to more carbon dioxide in the atmosphere (which leads to global warming). If you take away the forests, you also take away the carbon dioxide sinks. (A carbon sink is any reservoir that takes in carbon. Since plants use carbon dioxide for photosynthesis, forests are large carbon sinks that reduce the amount of carbon dioxide in the atmosphere. Other carbon sinks are oceans and soil.)

2. On the other side of the photosynthesis equation is the production of oxygen; deforestation leads to lower oxygen levels.
3. Decreased soil fertility.
4. Cutting down trees takes away habitat and food for many organisms, which means fewer species.
5. More flooding (plants collect moisture in roots and soil; if there are no plants to collect the moisture, it becomes run-off water that can lead to flooding).
6. Erosion occurs as sediments are transported with run-off water (since there are no plant roots helping to hold the soil together).
7. Erosion will lead to less clean water in surrounding areas as well.

There are many reasons for deforestation, but I will mention one that we can do something about. China makes 63 billion pairs of single-use chopsticks per year to use in China as well as to export to other countries. This results in the loss of 25 million poplar and birch trees every year. If these forests were left undisturbed, they would be a sink for 5 million tons of carbon dioxide, not to mention all of the animals that use these forests for their homes and all of the other benefits of forests listed above. Making trees into chopsticks also requires energy and causes immense CO_2 emissions. China is one of the world's largest polluters, with no real regulations in sight. Avoid using single-use chopsticks when possible and educate your family and friends about the issue as well. Bring your own reusable chopsticks to Chinese restaurants, or ask for a fork. If you get take-out be sure to decline the chopsticks that they usually put in the bag. Everything is based on supply and demand and if there is less demand, then hopefully we can curtail the use of disposable chopsticks in the future.

Deforestation causes the **extinction** of many species (all individuals of the species die, and so that species no longer exists). Numerous organisms are losing their habitats now. Every hour six species become extinct due to habitat loss. (This is the number-one cause of the mass extinction that is currently occurring.) **Extinctions** have always been a natural occurrence over the course of the history of life on Earth. Extinction and speciation rates have probably been pretty even until modern times. There have been millions of species on this planet, but 99% of them have become extinct over time. There have been periods when there were bursts of extinctions (verified with fossil evidence), and these punctuations are called "mass extinctions." A mass extinction means that 75% of the species go extinct within a relatively short time frame (300,000–2 million years). There have been six mass extinctions that we can document.

The first five historical reasons for mass extinctions were all due to natural causes:

1. Changes in sea levels.
2. Asteroids or meteors hitting the Earth (caused dinosaurs' extinction).
3. Climate change: Glaciations and warming periods.
4. Continental movements.
5. Habitat loss due to changes in land and sea.
6. Volcanoes—250 million years ago most of the marine life went extinct possibly due to volcanic eruption in Siberia.
   (Many of the mass extinctions were a combination of these factors.)

The sixth mass extinction, currently occurring, is directly related to human activity:

Habitat loss: This is the number-one threat and is affecting more than 90% of endangered species!

Introduced species: Exotic species are killing off the natives and destroying ecosystems.

Pollution: Greenhouse gases, smog, etc., causing **climate change**, acid rain, acidification, etc.

Population growth: There are 7 billion humans on our planet using up too many resources!

Overconsumption/overexploitation of wildlife is causing extinctions, especially in mammals and fish.

This 6th mass extinction is happening much quicker than the previous extinctions. Human activity is accelerating the number and pace of extinctions dramatically; in fact we could lose the majority of species on Earth now within the next 2,000 years if we don't curb our actions soon.

We need to:

Reduce consumption in order to curtail habitat loss, pollution and overexploitation and overconsumption. By reducing consumption we can also reduce waste (which will help reduce habitat loss and pollution). We produce so much trash that is not biodegradable that the areas for landfills are becoming scarce. (Every American discards about 100 TONS of garbage in a lifetime—that's the weight of a plane!) Burning trash causes air pollution, and shipping trash to the ocean causes water pollution. Some trash stays intact for generations, like plastics (12% of the garbage we discard is plastic). Plastic bags don't degrade for forty to fifty years! (So remember to bring your reusable bags to the grocery store.) My dad is an environmental lawyer and is helping to ban plastic bags in Southern California, which I'm very excited about. San Francisco has already banned plastic bags for most merchants thanks to former Mayor Gavin Newsom and California might even ban Styrofoam soon.

Try to use refillable bottles rather than plastic bottles for water as well, because all of the plastic bottles generated since bottled water became the rage are causing a huge problem. These bottles account for 1.5 million tons of plastic waste each year. Most are not recycled (only about 15% get recycled so 66 million bottles end up in the trash <u>every</u> <u>day</u>) and plastic bottles can take 1,000 years to decompose! There is a vortex of plastic waste in the Pacific Ocean called the "Pacific Garbage Patch" or the "North Pacific Gyre." It is a 5-million-square-mile island of floating plastic debris 1,000 miles from the California coast—twice the size of Texas! (Eighty percent of the pollution in the ocean in general is from plastic, Styrofoam, and cigarette butts.) Our plastic trash in the ocean is causing the death of countless fish, seabirds, sea turtles, and other marine life. If that weren't bad enough, the bottle industry uses 47 million gallons of oil to produce the bottled water per year! Producing the bottles also releases over 2.5 million tons of carbon dioxide a year into our atmosphere, adding to global warming. Bottled water companies are sucking the water from local communities and polluting their environment in the process. (Watch the 2010 documentary "Tapped" by Stephanie Soechtig for more information.) And the final irony is that it takes almost 2 gallons of water to produce one plastic water bottle! Yet another precious resource that is wasted in the process.

Drinking bottled water is so wrong in so many ways. If the environmental concerns don't get to you, and you don't mind wasting money on water that is often more contaminated than tap water (tap water is more strictly regulated for contaminants than bottled water), then please think about some other health concerns. Plastics can leach toxic chemicals into the water (or food) that they hold, especially when they are heated. (Don't microwave in plastic containers and don't drink from water bottles that have been sitting in a hot car.) The toxins from plastics may be causing increased rates of diabetes, reproductive health issues, heart disease, autism, and prostate and breast cancers. Especially avoid plastics with a 3 or 7 inside the triangle since these have bisphenol A (BPA) and phthalates, which are linked to endocrine developmental problems and can harm even at very low levels. Some countries have banned BPA from being used in baby bottles and the FDA is in the process of evaluating it for safety in other containers now as well. BPA is also in the linings of some canned foods.

If you really must buy water, then at least buy it in a can rather than in a plastic bottle. It takes less energy to produce cans, they are more easily recycled and cans don't leach as many toxic chemicals that plastics do. There is a great organization called cannedwater4kids.com that sells canned water and uses 95% of the proceeds to supply water to children around the world. My high school, Crossroads, stopped offering bottled water for sale on campus a few years ago and started ordering water through this company. Small changes like this can really help in the long run. An even better thing to do is not buy any water and just keep refilling a glass bottle from the tap—I use a Juice Squeeze bottle, which is the perfect size for me, but Snapple bottles are great too. You can clean them out and reuse them again and again for your water. (I said to my daughter Ella one day "I wish everyone knew not to use plastic water bottles" and she replied "me too, then I wouldn't be the only one at school drinking water out of a Juice Squeeze bottle!" My kids may be embarrassed by my crazy ways but at least they are healthy.

Reduce waste, compost, recycle, and reuse whenever possible. The energy saved by recycling 1 aluminum can is the equivalent of 3 hours of TV electricity or half a gallon of gas! Recycling composting and reusing not only saves important resources such as energy, water and money, but also saves landfill space, which is also critical so that we don't have to destroy more habitat to make new landfills. (Some states have excellent recycling programs but most of the country does not recycle enough.)

Conserve energy—turn off your lights, <u>shut down</u> your computer (don't just put it to sleep at night), use less gas by using mass transit, walking, biking, carpooling, look for energy-efficient appliances, use less hot water, get a hybrid or electric car, support green businesses, buy locally so that less energy is used to ship goods across the world, etc., etc.

Support renewable energy efforts—Sun, wind, water, and geothermal. It's becoming more convenient to install solar panels on the roof of your house now, which is exciting—PG&E even credits your account when you produce more energy than you consume.

Conserve water—much of the United States is experiencing water shortages, especially in California, Arizona, Hawaii, and Texas. Water issues are even worse in other drier areas around the world, and account for thousands of deaths every day—roughly 3.5 million a year. Water is important for drinking, food supply, sanitation, hygiene, and health facilities. You can live for weeks without food, but without water, you will perish within days. Heating water (for showers, etc.) uses energy as well. So try taking shorter showers, conserve water by brushing your teeth with the faucet turned off, water the garden in the afternoon rather than in the heat of the day, and install low-flush toilets (if you own a home). Suggest some of these practices to conserve water to your family and friends as well.

Let's go over some of the major threats to biodiversity in more detail.

Problems with overpopulation:

Population growth is reaching the <u>carrying capacity</u>. If we continue to grow the way we are, we won't have the amount of resources needed to sustain life on Earth. (We won't be able to grow enough food to feed everyone, finding clean water will be a problem, and we won't have the energy sources and raw materials to sustain the world population.) There are over 7 billion humans now on Earth. If we continue to increase our population growth at our current rate, there will be 12 billion in 10 years! Environmental recommendations are to only to replace yourself—don't have more than 2 children per couple in order to help curb our population growth. The U.S. is especially greedy in our resource consumption. We are only 5% of the world but we use 20% of the resources. If the rest of the world used as many resources as we do, we would need the resources of eight Earths!

Human expansion and resource consumption lead to more and more **pollution**, which not only harms human health but also affects all of the organisms in the polluted area. **Pollutants** are substances that harm health, activities, and survival of species and populations. We are constantly bombarded with toxins, from food (pesticides, colorants and preservatives), from water (toxic run-off from factories, energy extraction processes, chlorine, etc), and from toxic gases like carbon monoxide and ozone released into the air. Pollutants have different effects on our bodies. They can affect our circulatory, respiratory, endocrine, muscular, digestive, integument, immune, and nervous systems. Some common ailments from pollution include asthma and other respiratory disorders, weakened immune systems, headaches, arthritis, gastritis, nerve damage, heart disease, and cancer.

Health effects of pollution

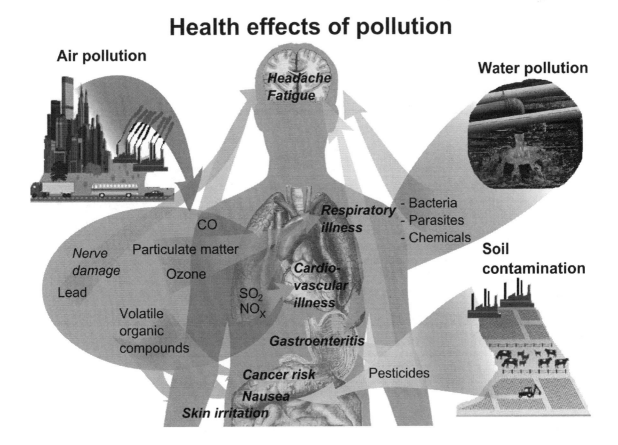

Human pollutants affect other organisms as well, and can destroy whole ecosystems. Water pollution is a huge issue and can be due to run-off from power plants, factories, and agriculture, in addition to community run-off, which can include pesticides, fertilizers, carcinogens, radioactive material, and solid waste from sewage and trash (especially from plastics, which are causing a huge sludge in different parts of the world). There is massive water pollution every time there is an oil spill as well. The spill in the Gulf of Mexico has killed countless organisms that used to thrive in that ecosystem. Water pollution can also occur from burning fossil fuels, which results in the dumping of massive quantities of carbon dioxide into the ocean (22 million tons every day!). Carbon dioxide pollution of the oceans is causing **acidification**, which is affecting marine animals like shellfish, corals, and plankton, which need to build shells for survival. The lower pH of the water with the addition of carbon dioxide is impairing their ability to build these shells. Many other species rely on these shelled organisms for food and protection. For example, the fish that live within coral reefs will be wiped out if the corals can't survive. Acidification, if left unchecked, will destroy whole ocean ecosystems, which in turn will affect the communities that depend on the coastal resources.

You can easily see the effects of water pollution, but some pollutants are harder to detect, and we are just now starting to understand their long-lasting effects.

Most of our Air Pollution is from burning fossil fuels, which we use for energy. **Fossil fuels** are ancient remains of plants and animals that decomposed millions of years ago, and then were buried, heated, and pressurized by the Earth's crust. Examples of fossil fuels include coal, oil, and natural gas. These are all non-renewable resources. We <u>will</u> run out of fossil fuels eventually—the only question is when. We are trying to move toward more renewable sources of energy such as solar, water, geothermal, tidal, and wind. We all need to conserve energy more as well: use less gas, turn off electricity when you don't need it, etc., so that we can slow down our consumption of energy and reduce our "carbon footprint."

Burning fossil fuels intensifies the greenhouse effect by increasing the amount of **greenhouse gases** (CO_2, ozone, methane, water vapor, and nitrous oxide) in the atmosphere. Greenhouse gases have occurred naturally throughout history and have a very important function of absorbing and emitting infrared radiation from the Sun. These gases accumulate in the Earth's atmosphere and are like a cloud that traps some of the Sun's heat, helping to keep the average temperature on Earth around 59°F, which is pleasant to life on Earth. (Otherwise more of the sun's radiation would be reflected from the Earth's surface and atmosphere, and the temperature on Earth would be about 1°F.) Human activity has drastically increased greenhouse gases, mostly by increasing CO_2 emissions from the combustion of fossil fuels for electricity and transportation. Increased CO_2 levels in the atmosphere causes global temperatures to rise as the greenhouse gas layer gets thicker and traps more heat in our atmosphere. Human activity has increased carbon dioxide levels by 30% over the naturally occurring levels of CO_2. This is creating "Global Warming," which is an accurate term to use when describing the overall increase in the temperature of the planet due to higher carbon dioxide levels. However, most people now use the term "Climate Change" due to the fact that the overall warming of our planet is causing many different weather patterns. (Increased temperatures causes ice to melt and water to evaporate and the increased water vapor in the atmosphere is moved around by winds to different areas and can come back to earth in the form of fog, rain, hail, sleet, snow, etc., so some areas are actually cooler than usual.) Climate change due to global warming is having many adverse affects on the planet. For example, summer arctic sea ice has decreased by more than 30% in the last 30 years and could be completely gone by the summer of 2040! I hate to think that there may be no more polar bears in the wild by the time I have grandchildren. Countless other species, including humans, are affected as well. Let's go over some of the affects here.

There are MANY very serious consequences of Global Warming/Climate Change:

1. Polar ice/Glaciers are melting—this affects our fresh water supply. Forty percent of the human population relies on glacier ice for their water! There are already water shortages in some areas.
2. Sea levels are rising as the ice is melting, causing problems for many species.
3. Warmer waters affect marine food webs. It starts with the phytoplankton and then continues up the food chain. The phytoplankton are blooming earlier than usual, which throws off the food supply for other organisms, like whales, that migrate to eat the phytoplankton.
4. Weather changes are not just heating up the Earth, they are causing other natural disasters including hurricanes, cyclones, tornadoes, tsunamis, blizzards, rain, and drought.
5. Higher evaporation of water on Earth means that lakes, reservoirs, etc., will dry up. There are already important reservoirs that have dried up completely.
6. Climate change is causing habitat loss and food scarcity; changing animal migration, hibernation, and mating cycles; and leading to many extinctions. It could be the leading cause of extinction now along with habitat destruction.

World leaders met in Kyoto in 1997 to draw up the Kyoto Treaty to call for worldwide emission controls in order to reduce the effects of global warming. Everyone signed it except for the United States—one of the top offenders! Scientists working on global warming at the time predicted that by 2100 the average temperature around the world could rise 3.6 to 8.1°F if our current consumption continued. Luckily, now there are more people who realize that global warming is an indisputable fact, but it's still slow going to get people to reduce their consumption of energy.

Other effects of burning fuels: Oxides, such as sulfur and nitrogen oxides released from factories and automobiles due to fossil fuel combustion, react with water in the atmosphere to form **acid rain**. This falls back to Earth as rain or snow and is very acidic, harming amphibians, fish, reptiles, and plants the most. Whole forests are dying due to acid rain in some areas.

Pollution is also causing the Ozone Shield to be depleted (mostly over Antarctica near southern Australia, but it is happening at all latitudes). The ozone shield screens out 99% of the Sun's ultraviolet radiation. There will be more incidences of skin cancer as this shield is destroyed. (People in Australia are already feeling the effects, with more skin cancer and second-degree burns!) Ozone Depleting Substances (ODS), such as compounds with chlorine and bromine, are the main pollutants causing this depletion. They break up ozone molecules when they are released into the stratosphere and leave behind oxygen, which does nothing to protect the Earth from ultraviolet radiation. Chlorofluorocarbons (CFCs) are one of the main culprits burning a hole in this shield. CFCs are used for refrigerants, aerosol propellants, and solvents. (The use of CFCs has been decreased since this was discovered to be the main cause of ozone depletion.)

Research in the past forty years has provided much information about how humans are impacting the Earth with our actions and pollution. Earth Day was started on April 22nd, 1970 in San Francisco, California and was declared a national holiday forty-two years later by President Barack Obama. It is now celebrated all over the United States as well as in 100 other countries. It is the most celebrated environmental event worldwide and has been very important for bringing awareness to so many environmental concerns. The United States Congress passed the Clean Air, Clean Water, Safe Drinking Water, Marine Mammal Protection, and Endangered Species Acts in the past 40 years, and founded the Environmental Protection Agency. Such environmental awareness, activism, agencies, and laws have helped to greatly improved air and water quality, bring stricter regulations for emissions, pollutants, and activities that harm other organisms, and promote more environmental education in schools. Hopefully we will continue to learn how to avoid the damages that we are causing now, move toward more renewable resources, and clean up the Earth for future populations of all species. You can do so much to help with conservation efforts, too. Every little bit helps: reuse bags, bus, carpool or bike to school, conserve water with shorter showers, don't buy bottled water, shop locally, and definitely educate others on these issues. Supporting and volunteering for environmental organizations and writing to politicians are also great ways to help with conservation efforts. It's critical to vote for politicians who protect the environment and our health as well. Supporting education and becoming educated yourself is very important—so congratulations on taking that first step! Understanding biology is crucial for helping the planet and our own health.

We've come full circle from the first chapter where I discussed why people study biology. I hope this book has provided you with a good grasp of how your body works, what you can do to keep it healthy, and how you can help the planet. I also hope that you will further research any of the topics that you found interesting, and that you will share some of the information with your family and friends. I have so much more to say, but I'll stop here except to remind you of some of the most important ways to **protect your health and the environment**:

Don't smoke.
Safe sex and healthy relationships.
Get your vaccinations and booster shots.
Exercise your body and brain.
Eat a healthy, balanced diet.
Don't get too stressed and get enough sleep.
Social interactions are beneficial.
Alcohol in moderation.
Be careful with your brain (no drugs).
Reduce carbon footprint, consumption, and waste (especially reduce plastics).
Conserve water and energy (take shorter showers, use less gas, electricity, etc.).
Recycle, compost, and reuse whenever possible.
Get your cancer screening tests.
Enjoy life (healthfully)!

Bibliography

Neil A. Campbell and Jane B. Reece, *Biology,* 7th edition (Menlo Park, CA: Benjamin Cummings, 2005).

Judith Goodenough and Betty A. McGuire, *Biology of Humans: Concepts, Applications, and Issues,* 3rd edition (Menlo Park, CA: Benjamin Cummings, 2010).

Kathleen A. Ireland, *Visualizing Human Biology,* 2nd edition (Hoboken: Wiley, National Geographic Society, 2010).

Rita Mary King, *Biology Made Simple* (New York: Broadway Books, 2003).

Beverly McMillan, *Human Body: A Visual Guide* (Buffalo: Firefly Books, 2006).

Cecie Starr and Beverly McMillan, *Human Biology,* 8th edition (Belmont, CA: Brooks/Cole, Cengage Learning, 2010).

Duff Wilson, "FDA to Examine Menthol Cigarettes," *NY Times,* March 29, 2010, B1.

Wilson, Plan to Widen Use of Statins Has Skeptics. *NY Times,* 31 March, 2010.

Stacey Colino, "Numbers to Live by," *Real Simple,* May 2010, 151.

Eleni N. Gage, "The Lowdown on Free Radicals," *Real Simple,* December 2009, 135.

Deb Schwartz, "Organizing for Your Personality," *Real Simple,* June 2010, 198.

"Preserving and Boosting Your Memory: 10 Steps to an Optimal Memory," Harvard Medical School, Harvard Health Publications: <www.health.harvard.edu>.

<http://www.prostate-massage-and-health.com/prostate-stimulation.html>.

<http://www.cytochemistry.net/cell-biology/nucleus3.htm>.

<http://nationalzoo.si.edu/Animals/Primates/Facts/Primateness/default.cfm>.

<http://www.cancer.org/>.

<http://www.mydr.com.au/gastrointestinal-health/liver-and-alcohol-breakdown>.

<http://www.sciencedaily.com/releases/2008/11/081126081403.htm>.

<http://en.wikipedia.org/wiki/Timeline_of_evolution>.

<http://www.earthlife.net/mammals/evolution.html>.

<http://www.ted.com/talks/william_li.html>.

<http://www.niams.nih.gov/health_info/bone/Osteoporosis/Conditions_Behaviors/default.asp>.

<http://kidshealth.org/kid/htbw/htbw_main_page.html>.

<http://depts.washington.edu/learncpr/>.

<https://health.google.com/health/ref/Gallstones>.

<http://www.mypyramid.gov>.

<http://www.caloriesperhour.com/tutorial_BMR.php>.

<http://news.bbc.co.uk/2/hi/health/7151813.stm>.

<http://www.nationmaster.com/graph/hea_obe-health-obesity>.

<http://www.cdc.gov/nchs/fastats/lcod.htm>.

<http://www.ultimatefatburner.com/hcg-review.html>.

<http://www.pbs.org/wgbh/nova/evolution/how-did-life-begin.html>.

<http://en.wikipedia.org/wiki/Abiogenesis>.

<http://health.yahoo.net/experts/healthieryou/tired-water-read>.

<http://health.yahoo.net/articles/womens-health/10-tricks-improving-your-memory>.

<http://online.prevention.com/oddbodyexplained/list/1.shtml>.

<http://en.wikipedia.org/wiki/Laughter>.

<http://dl.clackamas.edu/ch106-08/structur1.htm>.

<http://faculty.washington.edu/chudler/heshe.html>.

<http://www.healthstatus.com/calculate/fsz>.

<http://www.fi.edu/learn/heart/blood/platelet.html>.

<http://www.mayoclinic.com/health/fat/NU00262>.

<http://www.investorplace.com/20714/chocolate-consumption-coronary-heart-disease-risk/?sms_ss=email&at_xt=4cbe3d0f6ff60b5e,0>.

<http://en.wikipedia.org/wiki/Noncoding_DNA>.

Philip Whitfield, ed., *Human Body Explained: A Guide to Understanding the Incredible Living Machine*, Henry Holt Reference book (New York: Henry Holt and Company, 1995).

Body: The Complete Human: How it Grows, How It Works and How to Keep It Healthy and Strong (National Geographic Society: Washington DC, 2007).

Joel Robertson and Tom Monte, "Peak Performance" (Harper: San Francisco, 1996).

<www.health.harvard.edu/usda-protein>.

Celeste Robb-Nicholson, "By the Way Doctor," *Harvard Women's Health Watch* 18.3(2010): 8.

"The Botany of Desire," (PBS video based on the book by Michael Pollan, *The Botany of Desire: A Plant's Eye View of the World*).

David G. Myers, *Psychology in Everyday Life* (Hope College, Holland, MI: Worth Publishers, 2009).

Edward Rothstein, "Glimpsing the Brain's Powers (and Limitations): Interactive Show Explains How a Complex Organ Struggles to Make Sense of the World," *NY Times,* November 20, 2010, C1 & C5.

Jill Bolte Taylor, Donate Brain Tissue Shortage. National Alliance on Mental Illness, 22(Winter 2010).

Sandra Aamodt and Sam Wong, *Welcome to Your Brain* (New York: Bloomsbury, 2008).

Maureen Murdock, "Memoir on Contemporary Myth" (PhD diss., in Mythological Studies with an Emphasis in Depth Psychology, Pacifica Graduate Institute, 2010).

Elizabeth A. Phelps and Tali Sharot, "How (and Why) Emotion Enhances the Subjective Sense of Recollection" *Current Directions in Psychological Science* 17.2(2008): 147–152.

Stephen Jay Gould, "Male Nipples and Clitoral Ripples," *Bully for Brontosarus: Reflections in Natural History* (New York: W.W. Norton & Company, 1991).

<http://health.yahoo.net/experts/eatthis/8-best-supermarket-sweets>.

<http://www.youtube.com/watch?v=dBnniua6-oM>.

<http://www.sharecare.com/question/how-smoking-affects-facial-skin>.

Eric Wagner, "Whales in a Warming Sea," National Wildlife Federation, Aug/Sept 2011, 39.

"Mass Extinction May Have Been Caused By Volcanoes," *Science Illustrated,* Sept/Oct 2011, 16.

David Servan-Schreiber, *Anti-Cancer: A New Way of Life* (Viking, Penguin Group, 2008).

<bbc.co.uk/news/health-16031149>.

<http://en.wikipedia.org/wiki/Osteoarthritis>.

<http://www.ehow.com/about_5533277_many-bones-there-human.html>.

<http://www.nytimes.com/2012/04/22/magazine/how-exercise-could-lead-to-a-better-brain.html?pagewanted=1&_r=1&emc=eta1>.

<http://www.nytimes.com/2012/04/22/magazine/can-you-make-yourself-smarter.html?pagewanted=1&ref=general&src=me>.

<http://ornl.gov/sci/techresources/Human_Genome/home.shtml>.

<http://www.wired.com/medtech/stemcells/news/2008/01/blastocyst_biopsy>.

<http://info.kp.org/html/partnersinhealth_ncal/may2012/obesity.html>.

<http://info.kp.org/html/partnersinhealth_ncal/may2012/general_health.html>.

<http://www.wired.com/medtech/health/news/2005/01/66198>.

<http://cannedwater4kids.com/>.

<http://www.fda.gov>.

<http://iipdigital.usembassy.gov/st/english/texttrans/2012/04/201204204373.html#axzz1w1gRgom4>.

<http://climaterealityproject.org>.

Brian Walsh, "The Perils of Plastic," *Time Magazine*, April 2010.

Frank R. Spellman, *Biology for Nonbiologists* (Toronto, Canada: Government Institutes, 2007).

Marvin M. Lipman, *The Best of Health*, Consumer Reports on Health (Consumers Union of United States, Inc., 2011).

Index